Cytokines

Edited by Payam Behzadi

Published in London, United Kingdom

IntechOpen

Supporting open minds since 2005

Cytokines
http://dx.doi.org/10.5772/intechopen.77671
Edited by Payam Behzadi

Contributors
Xuehui He, Xinhui Wang, Mubarak Muhammad, Matjaž Jeras, Ariana Barlič, Sara Žigon-Branc, Harumi Jyonouchi, Dorin Gheorghe, Petra Surlin, Popescu Dora, Anne Marie Rauten, Alexandra Martu, Silvia Martu, Liliana Foia, Sorina Solomon, Vasilica Toma, Adrian Camen, Simona Popa, Allma Pitru, Ion Rogoveanu, Madalina Olteanu, Mihail Virgil Boldeanu, Misbahul Arfin, Abdulrahman Al-Asmari, Samuel Evans Adunyah, Richard Akomeah, Chantel F. Fuqua, Elisa García-Vences, Roxana Rodríguez-Barrera, Adrián Flores- Romero, Julián García-Sánchez, Lisset Karina Navarro-Torres, Marcela Garibay Lopéz, Payam Behzadi, Márió Gajdács

First published in London, United Kingdom, 2020 by IntechOpen
IntechOpen is the global imprint of INTECHOPEN LIMITED, registered in England and Wales, registration number: 11086078, 7th floor, 10 Lower Thames Street, London, EC3R 6AF, United Kingdom
Printed in Croatia

British Library Cataloguing-in-Publication Data
A catalogue record for this book is available from the British Library

Additional hard and PDF copies can be obtained from orders@intechopen.com

Cytokines
Edited by Payam Behzadi
p. cm.
Print ISBN 978-1-78984-858-8
Online ISBN 978-1-78984-859-5
eBook (PDF) ISBN 978-1-83968-339-8

For EU product safety concerns:
IN TECH d.o.o., Prolaz Marije Krucifikse Kozulić 3, 51000 Rijeka, Croatia, info@intechopen.com or visit our website at intechopen.com.

Meet the editor

Dr. Payam Behzadi was born in Tehran, Iran, in 1973. He has a BSc and MSc in Microbiology and a PhD in Molecular Biology. He is currently Assistant Professor in the Department of Microbiology, College of Basic Sciences, Shahr-e-Qods Branch, Islamic Azad University. He has authored and edited twenty chapters and academic books, and seventy original and review articles. His scientific research interests include urinary tract infections, bioinformatics, genetics, gene profiling, and molecular biology.
Dr. Behzadi trains as an ice skater in his free time.

Contents

Preface

The human body consists of a diversity of tissues, organs and systems that all perform their functions in accordance with their own inherited characteristics and properties. However, a stable harmony is needed for these structures to collaborate with each other as a united biological system.

In this regard, the immune system has a pivotal role to orchestrate, regulate and modulate all of the body's biological processes and systems. The immune system is a complicated two-part network that encompasses cells, molecules, proteins and glycoproteins. The building blocks of these components collaborate with each other as well as other structures in the body in a harmonic and stable pattern. Hence, the homeostasis of different systems and metabolic processes is provided by molecules recruited by the immune system.

Among a wide range of immunological recruited molecules, cytokines have a considerable position in this regard. Cytokines are involved in every immune and non-immune system and structure. Each cytokine is able to induce its own production as well as that of other cytokines and simultaneously activate the production of other molecules. Cytokines play their role in the forms of autocrine, paracrine and endocrine.

Interestingly, cytokines link to their specific receptors and activate the related signaling pathways. In this regard, cytokines recruit different cells and molecules from both the immune and non-immune systems. This process results in homeostasis.

This complicated network of cytokines is important for harmonizing between different processes and structures in the human body. As such, any mutations at the gene level of the related cells and molecules may lead to a wide range of disorders and diseases. Therefore, cytokines and related molecules can be recognized as important for treatment of immune and non-immune disorders and diseases.

Our knowledge regarding the importance of immune system goes back two millennia. Today, by the progression of cellular and molecular biology, bioinformatics, computational biology and chemistry, we know a wide range of immunological processes and mechanisms.

The book contains 9 chapters.

The introductory chapter provides a background about historical evidences regarding immunology, immune system and cytokines.

Chapter 2, "*In vitro* Cell-Based Assays for Potency Testing of Anti-TNF-α Biological Drugs"; Chapter 3, "Tumor Necrosis Factor Alpha: A Major Cytokine of Brain Neuroinflammation"; and Chapter 4, "TNFR2 and Regulatory T Cells: Potential Immune Checkpoint Target in Cancer Immunotherapy" focus on the role and importance of tumor necrosis factor (TNF) in human body.

Chapter 5, "Innate Immunity and Neuroinflammation in Neuropsychiatric Conditions Including Autism Spectrum Disorders: Role of Innate Immune Memory"; Chapter 6, "Cytokines in Scar Glial Formation after an Acute and Chronic Spinal Cord Injury"; Chapter 7, "The Genetic Aspects of Behçet's Disease: Role of Cytokine Genes Polymorphisms"; and Chapter 8, "IL-21 Signaling and Induction of Cytokine Expression in Human Leukemia Cells and Monocytes" present information about the association of cytokines and different disorders and diseases.

Finally, Chapter 9, "Cytokines' Involvement in Periodontal Changes," examines the role of cytokines in dentistry.

I appreciate all the authors who contributed to this book for their great information and collaboration.

I would like to give special thanks to my colleague Dr. Márió GAJDÁCS from Hungary, who collaborated with me to prepare the introductory chapter. I am also grateful to Author Service Managers Romina Skomersic and Dajana Pemac and Commissioning Editors Lucija Tomicic-Dromgool and Martina Usljebrka Kauric at IntechOpen for their excellent collaboration and management during the production of this valuable book.

Payam Behzadi
Assistant Professor,
Department of Microbiology,
College of Basic Sciences,
Shahr-e-Qods Branch,
Islamic Azad University,
Tehran, Iran

Chapter 1

Introductory Chapter: Cytokines - The Diamonds and Pearls of Biological Systems

Márió Gajdács and Payam Behzadi

1. The historical background of immunology

When we are reading or investigating the topics of cellular and molecular immunology including cytokines, our unconscious mind suddenly goes back to the Iranian "Immunologist King," Mithridates VI Eupator of Pontus. Pontus is a Greek term referring to the Sea; but conceptually – in accordance with geographic evidences of that time period – it refers to the Black Sea. Indeed, he is known as the first immunologist in the ancient world [1–4].

Mithridates VI Eupator of Pontus (reigned 120–163 BC), the son of Mithridates V Euergetes, was proud over his Iranian heritage and ruled the Pontus region (Asia Minor, The Great Persia Empire Region; Present-day Turkey) throughout the Hellenistic and Mithridatic Kingdoms. Despite accepting Hellenism, he always considered himself as an Iranian king from royal Achaemenid lineages. Indeed, his Iranian origin went back as far as Darius I and Cyrus the Great (The King of Kings). His name Mithridates depicts the God of Light (Mithradatha: sent (given) by Mithra, the ancient Iranian God of Sun) [1, 2, 5–7].

Mithridates VI Eupator of Pontus is the founder of the *Mithridatism* theory which refers to the phenomenon of acquired immunity against poisons, by using determined doses of the poison to expose the individual to the agent little by little [8]. Hence, he used cocktails of antidotes against currently known poisons every day. The potion of Mithridates VI Eupator of Pontus was a mixture of 54 currently known poisons which was termed "**Antidotum Mithridaticum.**" His popular poisonous potion was consumed by people from all walks of life for about two millennia [6, 9, 10]. This process is very similar to the act of vaccination and may be considered as a preliminary form of preventive medicine [11].

Mithridates VI Eupator of Pontus was a genius because he was not only an exceptional expert in immunization and toxicology, but he could also talk in 22–25 different languages. Moreover, he was interested in medicine and pharmacology; therefore, Mithridates VI Eupator of Pontus has written several treatises regarding the characteristics of "**materia medica**" together with the related cases. These invaluable Mithridatic treatises were translated into Latin between 95 and 25 BC by Lenaeus, which were recognized as unique and effective prescriptions in Rome [12].

He was a researcher in the fields of immunology, toxicology, pharmacology, and medicine [4, 11, 12]. Due to this fact, the Iranian king, Mithridates VI Eupator of Pontus is a shining star in the treasure of science and history of Iran.

 IntechOpen

2. The immune system and cytokines

After about 2000 years, the basis of our knowledge and understanding of the immune system and immunity is very similar; however, what has changed is our ability to have an invaluable interpretation from our scientific observations.

The progression of cellular and molecular biology, together with bioinformatics, computational biology, and medicinal chemistry has given us an opportunity to have an effective understanding of the role of versatile molecules, proteins, and glycoproteins in different fields of immunology.

As we know today, the immune system is a complex network, containing wide range of cells and molecules which works rigorously and around the clock. The main sections of the immune system are divided into the innate and adaptive immune structures, while at the same time they create a unite and interdependent complex.

The unity of innate and adaptive immune system is supported by the diversity of common molecules, cells, mechanisms, processes, and pathways. However, the innate immune system is activated first and by the continuous of the presence of the unknown antigen(s), the adaptive immune system will be activated, usually a few days later. Besides, B- and T- lymphocytes have pivotal roles in maintaining the adaptive immune system and a long-lasting immune response [13].

The immune system – whether innate or adaptive– employs versatile mechanisms to protect the host's body from various "invaders," including pathogens or any other agents which may cause any type of disease. In this regard, the proteins of cytokines and chemokines are produced by different activated immune and non-immune cells and cell receptors, which they convoke a mass of molecules and cells into the center of infection [13].

Therefore, cytokines are noteworthy molecules, with a diversity of activities, functions, structures, and potential abilities. Due to this characteristic of cytokines makes them like "diamonds" and "pearls" of the immune system. Cytokines are invaluable treasure of the immune system with a high plasticity in functions and structures.

They join not only the all parts of the immune system together and unite them as a whole, but also the cytokines contribute in non-immune cells and molecules to orchestrate different cells, tissues, organs, and systems. That is why we may call cytokines as diamonds and pearls of the biological systems.

3. Immune system and cytokines

Cytokines were first described as soluble pyrexins by Menkin in 1944 [14]. The term "cytokine" was proposed (coined) by Cohen et al. through their commentary, published in 1974 [15, 16].

From that point onward, our knowledge regarding cytokines increased significantly. Therefore, in 1978, more than 100 different types activities were described, in association with cytokines [14].

Cytokines comprise a diverse group of proteins and glycoproteins, with molecular weights ranging between 5 and 20 kDa [17, 18]. These amazing molecules contribute to different biological and physiological mechanisms, such as blood pressure, inflammation, and cellular metabolism [18, 19]. Cytokines act locally, and therefore, they affect their peripheral cells in a paracrine style and the producing cells in an autocrine manner. Those cytokines, which disseminate via blood stream across the body act in an endocrine fashion, similarly to hormones. Due to this fact, there are no clear differences between hormones and cytokines, and the employed molecules by different biologic systems involving endocrine functions, hematopoiesis, immunity, and the nerves have similar characteristics in their functions and structures [13, 14, 16]. In another words, cytokines orchestrate hematopoietic, immune, and non-immune cells of the host in their development, differentiation, function, growth, and regulation, throughout the employment of paracrine and autocrine signaling pathways [13, 16, 19, 20]. Indeed, cytokines regulate, modulate, and orchestrate biological systems, for example, innate and adaptive immune system networks by induction of their own secretion [13]. In this regard, the Human Genome Project (HGP) has had an important role to recognize the different types of cytokines and their association with health and diseases [19]. The homeostasis across the systems of the human body is provided by an influent balance between anti-inflammatory and pro-inflammatory cytokines. Hence, the expression of genes producing cytokines is entirely modulated by long non-coding RNAs, both in transcriptional and post-transcriptional phases [18].

Cytokines are multifunctional proteins with a versatile of receptors, which are distributed in different systems of the body. Therefore, they can be categorized in accordance with variety plethora of criteria. However, it is recommended that the cytokines' classification would be on the basis of their receptors [16, 21]. Cytokines activate the related signaling pathway by their specific intracellular signaling cascades [13]. The major superfamilies of cytokine receptors are consisting of Transforming Growth Factor-β (TGF-β), Tumor Necrosis Factor (TNF) receptor, Serine Kinases family, Receptor Tyrosine kinases, Interleukin-1 (IL-1) and the related Toll-like Receptors (TLR), IL-17 receptors Type II (interferon), and Type I (hematopoietin) [16]. The chemokine receptor includes the G protein-coupled receptor family [19].

The type I receptor family, which is known as the hematopoietin receptor family, receives different types of cytokines including IL-2, IL-3, IL-4, IL-5 IL-6, IL-7, IL-9, IL-11, IL-12, IL-13, IL-15, IL-21, IL-23, IL-27, IL-31, IL-35, growth hormone (GH), prolactin (PRL), erythropoietin (EPO), thrombopoietin (TPO), leptin, granulocyte colony stimulating factor (G-CSF), ciliary neurotrophic factor (CNTF), leukemia inhibitory factor (LIF), oncostatin M (OM), cardiotropin-1 (CT-1), granulocyte macrophage (MΦ) colony stimulating factor (GM-CSF), and thymic stromal lymphopoietin (TSLP) [16]. Type I receptor family is structurally composed of homo- or heterodimers [19].

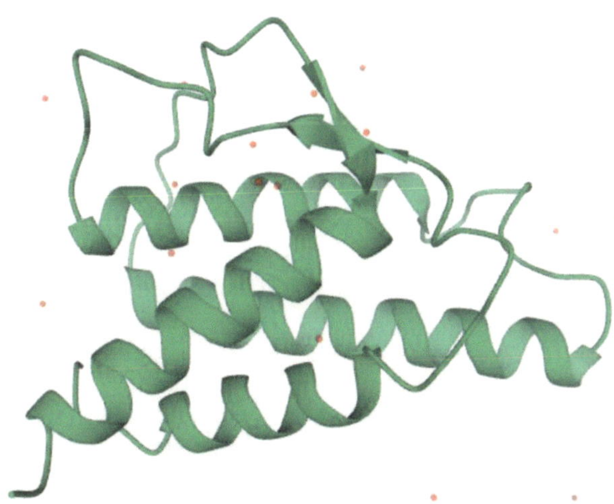

Figure 1.
The structure of IL-4 in details 1HIK PDB file [22, 23].

The type I cytokines are similar in their protein structure, with four anti-parallel α-helices with a configuration of up-up and down-down arrangement. The helices are connected with two long loops and one short loop [16]. This structure is shown in IL-4 (**Figure 1**) [22].

Type II receptor family or interferons (IFNs) with their homo- and heterodimeric structures have high affinity for their ligands, including IFN-α/β, IFN-γ, IL-10, IL-19, IL-20, IL-22, IL-24, IL-26, IL-28, IL-29, and IL-30 [16, 19].

The single pass membrane receptors of IL-1/TLRs bind to cytokines of IL-1α/β, IL-18, IL-33, IL-36, IL-37, and IL-38 [16, 19].

The IL-17 and TGF-β receptor serine kinase families bind to IL-7 and TGF-β cytokines, respectively [16].

The receptor tyrosine kinases, which phosphorylate tyrosine residues bind to a wide range of cytokines including IL-3, IL-16, IL-32, stem cell factor (SCF) and CSF-1, and FMS-like tyrosine kinase 3 (FLT-3) ligand [16, 19].

4. Cytokines and signaling pathways

Cytokines originate from the flat sac organelles of the Golgi-apparatus. These biological diamonds and pearls may have four different predestinations, including soluble cytokines released from endoplasmic reticulum, cytokines linked to the plasma membranes, intracellular cytosolic cytokines, and nuclear cytokines, and the latter group controls the process of transcription [19].

Regulation of cytokines may occur in different pathways, such as transcriptional and post-transcriptional regulations. At transcriptional level, the presence of several transcription factors determines the level of cytokine expression and production. In contrast to transcriptional regulation, the post-transcriptional regulation determines the duration of cytokines' expression. Moreover, the type of cytokines' glycosylation determines their functions and activity [18, 19, 24, 25].

As mentioned previously, there are several types of receptors which bind secreted cytokines. Due to this fact, the related receptors based on their signaling pathway are divided into four receptor groups. The first receptor group is in association with nuclear factor (NF)-κB and mitogen-activated protein kinases (MAPK); the second group involves receptors which employs Smad-family transcription factors; the third category of receptors activates the Ras extracellular signal-regulated kinase (ERK) pathway; and the fourth group which involves the majority of receptors uses the Janus kinase and signal transducers and activators of transcription (JAK-STAT) pathway [26].

The JAK family is composed of four members including JAK1, JAK2, JAK3, and TYK2, which are known as non-receptor multi-domain tyrosine kinases and recognized within the cell's cytoplasm [27, 28]. On the other hand, activation of JAK family members may lead to phosphorylation of tyrosine residues which belong to intracellular domains of the cytokine receptor. The STAT links to phosphorylated tyrosine residues of cytokine receptors. By STAT phosphorylation, the STAT allows for the ability of dimerization, transmission into the nucleus. Furthermore, the STAT will be able to manage the process of gene regulation [27–29].

These interactions reveal the importance of different cytokines, cytokines receptors, and cytokine signaling pathways; in which, the occurrence of any mutation may lead to different immune and non-immune disorders and diseases. Hence, in accordance with complicated interactions and communications in cytokines network, we can understand that recognition of cytokines, cytokine receptors, and their signaling pathways in details may lead us to a great opportunity for definite treatment of the immune and non-immune disorders.

We can conclude that, although Mithridates VI Eupator of Pontus as an Iranian Immunologist/Toxicologist/Pharmacologist King never had an idea about these biological "diamonds" and "pearls" regarded as cytokines, their receptors, and signaling pathway, he was aware of the harmonic communications between organs and systems within the human body.

Conflict of interest

The authors declare no conflict of interest, monetary or otherwise. The authors alone are responsible for the content and writing of this article.

Author details

Márió Gajdács[1,2] and Payam Behzadi[3]*

1 Department of Pharmacodynamics and Biopharmacy, Faculty of Pharmacy, University of Szeged, Szeged, Hungary

2 Institute of Medical Microbiology, Faculty of Medicine, Semmelweis University, Budapest, Hungary

3 Department of Microbiology, College of Basic Sciences, Shahr-e-Qods Branch, Islamic Azad University, Tehran, Iran

*Address all correspondence to: behzadipayam@yahoo.com

References

[1] Behzadi E, Behzadi P. Mithridates VI Eupator of Pontus: An ancient Iranian immunologist king. Central European Journal of Urology. 2017;**70**(3):323

[2] PONTUS. Encyclopædia Iranica [Internet]. 2004. Available from: http://www.iranicaonline.org/articles/pontus [Accessed: 20 September 2016]

[3] Behzadi E, Behzadi P. The role of toll-like receptors (TLRs) in urinary tract infections (UTIs). Central European Journal of Urology. 2016;**69**(4):404

[4] Behzadi E, Behzadi P. Basics of Laboratory Safety. 1st ed. Tehran-Iran: Niktab; 2008

[5] Mayor A. The Poison King: The Life and Legend of Mithradates, Rome's Deadliest Enemy. United States of America: Princeton University Press; 2009

[6] Jarcho S. Medical numismatic notes. VII. Mithridates IV. Bulletin of the New York Academy of Medicine. 1972;**48**(8):1059

[7] Pataci S, Lafli E. Archaeology of the southern Black Sea area during the period of Mithridates VI Eupator. In: Recent Studies on the Archaeology of Anatolia. Oxford: Archaeopress; 2015. pp. 313-325

[8] Grignolio A. The past, present and future of vaccines. In: Grignolio A, editor. Vaccines: Are They Worth a Shot? Cham, Switzerland: Copernicus Books: Springer; 2018. pp. 117-118

[9] Burcham PC. The emergence of modern toxicology. In: An Introduction to Toxicology. Springer-Verlag, London: Springer; 2014. pp. 1-27

[10] Yun J, Finkel T. Mitohormesis. Cell Metabolism. 2014;**19**(5):757-766

[11] Wexler P. History of Toxicology and Environmental Health: Toxicology in Antiquity II. London: Academic Press; 2014

[12] Totelin L. Mithradates' antidote–a pharmacological ghost. Early Science and Medicine. 2004;**9**(1):1-19

[13] Masi A, Glozier N, Dale R, Guastella AJ. The immune system, cytokines, and biomarkers in autism spectrum disorder. Neuroscience Bulletin. 2017;**33**(2):194-204

[14] Oppenheim JJ. Cytokines, their receptors and signals. In: The Autoimmune Diseases. London: Elsevier; 2020. pp. 275-289

[15] Cohen S, Bigazzi PE, Yoshida T. Similarities of T cell function in cell-mediated immunity and antibody production. Cellular Immunology. 1974;**12**(1):150-159

[16] O'Shea JJ, Gadina M, Siegel RM. Cytokines and cytokine receptors. In: Clinical immunology. London: Elsevier; 2019. pp. 127-155

[17] Behzadi P, Behzadi E, Ranjbar R. IL-12 family cytokines: General characteristics, pathogenic microorganisms, receptors, and signalling pathways. Acta Microbiologica et Immunologica Hungarica. 2016;**63**(1):1-25

[18] Carpenter S, Fitzgerald KA. Cytokines and long noncoding RNAs. Cold Spring Harbor Perspectives in Biology. 2018;**10**(6):a028589

[19] McInnes IB. Cytokines. In: Firestein GS, Budd RC, Gabriel SE, IB MI, O'Dell JR, editors. Kelley and Firestein's Textbook of Rheumatology. 10th ed. Philadelphia, PA: Elsevier; 2017. pp. 396-407

[20] Simpson S, Kaislasuo J, Guller S, Pal L. Thermal stability of cytokines: A review. Cytokine. 2020;**125**:154829

[21] Vilček J. The cytokines: An overview. In: The Cytokine Handbook. London: Elsevier; 2003. pp. 3-18

[22] Müller T, Oehlenschläger F, Buehner M. Human interleukin-4 and variant R88Q: Phasing X-ray diffraction data by molecular replacement using X-ray and nuclear magnetic resonance models. Journal of Molecular Biology. 1995;**247**(2):360-372

[23] Sehnal D, Rose A, Koča J, Burley S, Velankar S, editors. Mol* towards a common library and tools for web molecular graphics. In: Proceedings of the Workshop on Molecular Graphics and Visual Analysis of Molecular Data. Brno, Czech Republic: The Eurographics Association; 2018

[24] Liu J, Qian C, Cao X. Post-translational modification control of innate immunity. Immunity. 2016;**45**(1):15-30

[25] Arey BJ. The Role of Glycosylation in Receptor Signaling. London: IntechOpen; 2012

[26] Yoshimura A, Ito M, Chikuma S, Akanuma T, Nakatsukasa H. Negative regulation of cytokine signaling in immunity. Cold Spring Harbor Perspectives in Biology. 2018;**10**(7):a028571

[27] Ferrao R, Lupardus PJ. The Janus kinase (JAK) FERM and SH2 domains: Bringing specificity to JAK–receptor interactions. Frontiers in Endocrinology. 2017;**8**:71

[28] Ghoreschi K, Laurence A, O'Shea JJ. Janus kinases in immune cell signaling. Immunological Reviews. 2009;**228**(1):273-287

[29] Gajdács M, Spengler G, Sanmartín C, Marć MA, Handzlik J, Domínguez-Álvarez E. Selenoesters and selenoanhydrides as novel multidrug resistance reversing agents: A confirmation study in a colon cancer MDR cell line. Bioorganic & Medicinal Chemistry Letters. 2017;**27**(4):797-802

Chapter 2

In vitro Cell-Based Assays for Potency Testing of Anti-TNF-α Biological Drugs

Sara Žigon-Branc, Ariana Barlič and Matjaž Jeras

Abstract

Human cell-based assays for *in vitro* testing of drugs in preclinical and research studies, as well as in clinical practice, are gaining greater importance especially in view of personalized medicine, which is tailored to the individual needs and benefits of a patient. This chapter begins with an overview of contemporary cell-based assays, routinely used for a comparative *in vitro* potency testing of anti-TNF-α innovator biologics and their biosimilars. In sequel, based on the results of our original work, we will further discuss the establishment and use of 2D normal and osteoarthritic primary chondrocyte monolayer cultures and 3D microspheroidal articular cartilage tissues, prepared in hanging drops from osteoarthritic chondrocytes and chondrogenically differentiated mesenchymal stem cells. Both 2D and 3D cultures will be presented as models for assessing the neutralizing potency of the three well-known anti-TNF-α biological drugs: adalimumab, etanercept, and infliximab.

Keywords: *in vitro* cell-based assays, anti-TNF-α biologics, human articular chondrocytes, mesenchymal stem cells, 2D monolayer cultures, 3D cell cultures, gene expression

1. Introduction

Following the discovery and characterization of tumor necrosis factor (TNF) in the mid-1980s, this pleiotropic proinflammatory cytokine continues to be the focus of numerous studies and represents an important therapeutic target [1, 2]. The venue of anti-TNF biological drugs has revolutionized treatment of autoimmune and inflammatory diseases like rheumatoid arthritis, juvenile idiopathic arthritis, ankylosing spondylitis, psoriasis, Crohn's disease, ulcerative colitis, and others [2]. Although expensive, biological drugs (biologics) at the moment represent the best-selling group of pharmaceuticals. Nowadays, following the expiry of originators patents, a plethora of less expensive biosimilar drugs (biosimilars) are available to patients. In order to confirm the biocomparability of original and biosimilar products and to prove their quality, safety, and efficacy, the use of reliable and standardized bioassays relevant in assessing their modes of action is of crucial importance.

In this chapter, after a short introductory review of TNF biology, anti-TNF biological drugs and their mechanisms of action, we will present a selection of *in vitro* cell-based tests used either for general or personalized potency testing of anti-TNF biologics and their biosimilars.

2. A short overview of TNF biology

TNF is produced in various cell types, mainly immune cells such as monocytes and macrophages, microglia, neutrophils, natural killer cells (NK), T lymphocytes, and also in neuronal cells, keratinocytes, and fibroblasts [2, 3]. The cytokine exists in two biologically active forms. The first being a transmembrane protein (tmTNF), which can be cleaved by the metalloproteinase TNF-α-converting enzyme (TACE) (also known as disintegrin and metalloproteinase domain-containing protein 17 (ADAM17)) into its second form, a homotrimeric soluble TNF (sTNF) [2].

There are two TNF-binding homotrimeric transmembrane receptors, namely the TNF receptor 1 (TNFR1 or CD120a) and the TNF receptor 2 (TNFR2 or CD120b) [2]. While the TNFR1 is constitutively expressed on a vast majority of nucleated cells, the TNFR2 expression is inducible and tightly regulated, preferentially on endothelial, hematopoietic, neural, and immune cells [2, 4]. TNFR2 is also expressed on tumor cells where it is supposed to function as a tumor oncogene [5, 6].

Interestingly, tmTNF can induce signals in a bipolar way, as it acts as a ligand of both receptor types and as a receptor itself in cell-to-cell contacts [2, 4]. This means that tmTNF-α-expressing cells transmit signals to cells bearing TNFR1 and/or TNFR2. This phenomenon is called "outside-to-inside" or "reverse signaling," the function of which has not been completely clarified yet [2, 4]. The receptor function of tmTNF has been demonstrated in human monocytes, macrophages, NK cells, and T lymphocytes [4].

While TNFR1 is activated by both tmTNF and sTNF, TNFR2 can only be triggered by tmTNF. Both types of membrane-bound receptors are prone to TACE cleavage, resulting in fragments termed soluble TNF receptors (sTNFR) [2]. In turn, sTNFR may contribute to the regulation of cellular TNF responses by capturing and neutralizing circulating TNF (intrinsic TNF inhibitors). Additionally, due to increased receptor shedding, the number of functional signaling membrane TNFRs decreases. Consequently, this leads to a state of transient TNF desensitization [2].

3. Anti-TNF biological drugs and their mechanisms of action

Among currently available Food and Drug Administration (FDA)- and European Medicines Agency (EMA)-approved originator and biosimilar anti-TNF drugs, there are three full-length monoclonal antibodies (mAbs); these are infliximab (IFX), a chimeric mouse/human mAb (Remicade® and its biosimilars: Remsima®, Inflectra®, Flixabi®, Ixifi®, Renflexis®, and Zessly®), adalimumab (ADA), a fully humanized mAb (Humira® and its biosimilars: Cyltezo®, Imraldi®, Amgevita®, Solymbic®, Hyrimoz®, Hulio®, Halimatoz®, and Heyifa®), and golimumab, another fully humanized mAb (Simponi®) (**Figure 1**) [2, 4]. The additional two anti-TNF biological drugs, which are not mAbs, are etanercept (ETA) (Enbrel® and its biosimilars: Erelzi® and Benepali®), a fusion protein consisting of two extracellular parts of the human TNFR2 and the Fc portion of human IgG1, and certolizumab pegol (Cimzia®) composed of a human Fab' fragment, covalently attached to two cross-linked 20 kDa polyethylene glycol chains (**Figure 1**) [2, 4].

Although all anti-TNF biologics neutralize the same target (sTNF and tmTNF), they are not equally effective in treatment of certain inflammatory pathologies, for example, Crohn's disease. This is due to differences in their characteristics (structure and binding affinities) and mechanisms of action (**Figure 1**) [2, 4]. Besides all of them being efficacious in neutralizing both forms of TNF, infliximab additionally induces "outside-to-inside" signaling via binding to tmTNF, thereby triggering apoptosis of tmTNF-expressing immune cells [2]. Being full-length mAbs,

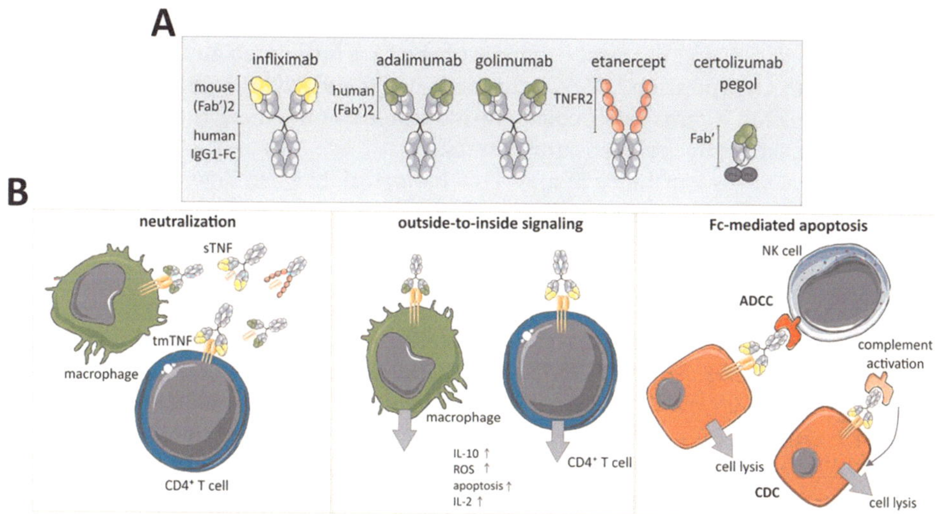

Figure 1.
Structures (A) and mechanisms of action (B) of the five FDA- and EMA-approved anti-TNF biologics; original figure used with the authors' permission under the terms and conditions of the Creative Commons Attribution (CC BY) license (http://creativecommons.org/licenses/by/4.0/) [2].

adalimumab, golimumab, and infliximab can, after binding to cells expressing tmTNF via their effector Fc regions (IgG1), induce antibody-dependent cytotoxicity (ADCC) of NK cells and activate the classical complement pathway, resulting in a complement-dependent cytotoxicity (CDC) and apoptosis [2, 4]. While ADCC and CDC are also induced by etanercept, which contains a truncated form of IgG1 Fc domain (lacking a CH1 constant region), certolizumab pegol, due to its Fc domain missing structure, acts differently. In treating inflammatory bowel disease with anti-TNF mAbs, another mechanism of their action is based on the interaction between IgG1 Fc domains of therapeutic mAbs and macrophage Fcγ receptors (FcγR), resulting in increased numbers of regulatory M2 macrophages ($CD206^+$). These cells in turn inhibit T cell proliferation [2, 7]. Additionally, in rheumatoid arthritis (RA), adalimumab enhances the expression of tmTNF on monocytes, thereby promoting the interaction between tmTNF and TNFR2 present on regulatory T cells (Tregs), which subsequently increase their immunosuppressive activities [2, 8]. Also in RA, infliximab promotes the generation of natural Tregs ($CD4^+CD25^{high}FoxP3^+$), which inhibit a proinflammatory cytokine production and replenish a defective pool of these cells, typically found in this autoimmune disease [2, 9]. In RA patients, the adhesion molecules and chemokines are upregulated on their joint vasculature endothelium. Blockage of TNF-α with adalimumab, golimumab, infliximab, or etanercept deactivates inflamed vascular endothelium, thereby decreasing the numbers of inflammatory immune cells entering synovial joints and additionally improving the generation of new synovial blood vessels by increasing the circulating levels of vascular endothelial growth factor (VEGF) [10, 11].

4. *In vitro* cell-based bioassays for general potency assessment of anti-TNF biologics

Numerous well-established and standardized cell-based assays are available for assessing and comparing potencies of anti-TNF biologics and their biosimilars. **Table 1** contains some basic information regarding the most frequently used routine TNF-α neutralization (A), ADCC (B), and CDC (C) tests. The majority of data on bioassays presented in **Table 1** (see next page) were summarized from two publications describing the establishment of the first infliximab and etanercept World Health Organization (WHO) International Standards [12, 13]. These were performed within international collaborative studies, confirming their high degree of relevance and analytical laboratory utility.

Additionally, the capability of anti-TNF biological drugs to downregulate E-selectin adhesion molecules expressed on inflamed vascular endothelium can be determined on *in vitro*-cultured human umbilical vein endothelial cells by using appropriately labeled anti-E-selectin detection antibodies [10].

Other bioassay readout approaches, like flow cytometry and measurement of induced endogenous gene expression by quantitative reverse transcription polymerase chain reaction (qRT-PCR), are also being applied [16–18].

The reason why various human and murine cell lines are used in these assays is that such tests can be standardized and their results can be compared between laboratories. However, the use of different types of primary cells in such general tests is less appropriate due to their high interindividual differences and in certain cases also weak responsiveness to anti-TNF biologics. Therefore, the results obtained in this way can hardly be compared [10].

5. Two-dimensional (2D) and three-dimensional (3D) primary cell cultures for personalized *in vitro* potency testing of anti-TNF biologics

Primary cells are indispensable for determining personal responses of patients to a given anti-TNF biologic, thereby generating important information for planning and performing optimal and cost-effective therapies. For this purpose, different cell types, especially those isolated from a patient's disease-affected tissues or *in vitro* differentiated autologous stem cells, can be used. In general, it is well established that in comparison to cells grown in 2D, those cultured in a 3D environment better mimic the scenarios *in vivo*. A number of cellular processes, that is, proliferation, differentiation, morphology, gene, and protein expressions, as well as responsiveness to external stimuli, are significantly affected by the physical aspects of the 3D environment [19–22].

In 2D cell cultures (monolayers), nutrients are evenly accessible to cells, but the communication between cells via secreted soluble molecules is restricted to their diffusion within the fluid, unless the medium is mixed or stirred regularly [23]. On the other hand, in dense multicellular 3D cell constructs prepared and cultured *in vitro*, nutrients and other soluble molecules, as well as oxygen supply, are limited by the mass transport, which is restricted by the construct's thickness/diameter and cell density per volume [23].

In the following subchapters, we will present our results after establishing 2D and 3D *in vitro* models for potency testing of anti-TNF biologics by using primary normal chondrocytes (NCs) and osteoarthritic chondrocytes (OACs), as well as chondrogenically differentiated bone-marrow-derived MSCs obtained from OA patients, with qRT-PCR gene expression assessment and protein secretion readout measurements.

5.1 Establishment of a 2D primary human chondrocyte-based cell model for *in vitro* testing of anti-TNF-α biologicals

Cartilage, which covers joint surfaces, is one of the most affected tissues in RA and other inflammatory arthritic diseases. Its only living constituents are chondrocytes, which produce and maintain a cartilaginous matrix mainly consisting

A

Cell types	rhTNF-α	Assay	Readout	Ref.
L929	10-86 IU/ml 7.2-20 IU/ml	Cytotoxicity	*Absorbance*: MTS, CCK-8 *Fluorescence*: Resazurin *Luminescence*: CellTiter-Glo®	[12] [13]
WEHI-164	100 IU/ml 2-100 IU/ml	Cytotoxicity	*Absorbance*: MTS, CCK-8/WST-8	[12] [13]
WEHI-13 VAR	5 IU/ml	Cytotoxicity	*Absorbance*: MTS, CCK8	[12]
KD4 Cl21	4.3 IU/ml	Cytotoxicity	*Absorbance*: MTS	[12]
KYM-1D4	0.1 ng/ml	Cytotoxicity	*Absorbance*: MTT	[10]
KYM-1D4	htmTNF-α expressed on mouse K2 cells	Cytotoxicity	*Absorbance*: MTT	[10]
U937	20-151.5 IU/ml 40-60 IU/ml 40 ng/ml	Apoptosis	*Luminescence*: Caspase-Glo® 3/7	[12] [13] [14]
HEK 293	17.2 IU/ml 4 ng/ml	Reporter gene	*Luminescence*: Steady-Glo®	[13] [14]
KJL	40 IU/ml	Reporter gene	*Luminescence*: Steadylite plus™	[12] [13] [15]

B

Effector cells	Target cells	E:T ratio	Readout	Ref.
NK92 expressing human Fc γ RIIIa	3T3 expressing htmTNF-α	1:1	*Luminescence*: CytoTox-Glo®	[12] [13]
Jurkat expressing human Fc γ RIIIa and NFAT-induced luciferase reporter gene	CHO-K1 expressing htmTNF-α	4:1	*Luminescence*: Bio-Glo™	[12] [13]
Jurkat expressing human Fc γ RIIIa linked to a luciferase reporter gene	HEK-239 expressing htmTNF-α	6:1	*Luminescence*: Dual-Glo®	[12] [13]
NK3.3	HEK-239 expressing htmTNF-α	10:1	*Fluorescence*: Calcein	[14]

C

Target cells	Complement	Assay	Readout	Ref.
Jurkat expressing htmTNF-α	Human serum	Viability	*Luminescence*: CytoTox-Glo®	[12] [13]
Jurkat expressing htmTNF-α	Human serum	Viability	*Absorbance*: CCK-8	[12] [13]
Jurkat expressing htmTNF-α	Human serum	Viability	*Luminescence*: CellTiter-Glo®	[14]

rhTNF-α—recombinant human TNF-α. Cells → CHO-K1: Chinese hamster ovary cells expressing human transmembrane TNF-α (htmTNF-α); HEK 293: human embryonic kidney cell line, transfected with the TNF-α-responsive NFκB-regulated Firefly luciferase reporter gene construct or expressing htmTNF-α; Jurkat: human acute T cell leukemia lymphocytes expressing htmTNF-α, resistant to TACE cleavage, human Fcγ RIIIa or TNF-α-responsive nuclear factor of activated T cells (NFAT) transcription factor-regulated Firefly luciferase reporter gene construct; K2: murine cells expressing the uncleavable htmTNF-α; KD4 Cl21: human rhabdomyosarcoma cell line; KJL: human erythroleukemic K562 cells transfected with the TNF-α-responsive NFκB-regulated Firefly luciferase reporter gene construct, together with the Renilla luciferase reporter gene under the control of a constitutive minimal thymidine kinase promoter; L929: murine fibroblast cell line; NK3.3: human natural killer (NK) cell line cloned from peripheral blood; NK92: NK lymphoblast cells from a malignant non-Hodgkin's lymphoma patient, expressing Fcγ RIIIa; 3T3: murine embryonic fibroblasts expressing htmTNF-α; U937: human histiocytic lymphoma cell line; WEHI-13 VAR and WEHI-164: murine rhabdomyosarcoma cell lines. Readout reagents (absorbance) → CCK-8/ WST-8: 2-(2-methoxy-4-nitrophenyl)-3-(4-nitrophenyl)-5-(2,4-disulfophenyl)-2H-tetrazolium, monosodium salt; MTT: 3-(4,5-dimethylthiazol-2-yl)-2,5-diphenyltetrazolium bromide; MTS: 3-(4,5-dimethylthiazol-2-yl)-5-(3-carboxymethoxyphenyl)-2-(4-sulfophenyl)-2H-tetrazolium, inner salt; luminescence detection kits: Bio-Glo™, Caspase-Glo® 3/7, CellTiter-Glo®, CytoTox-Glo®, Dual-Glo®, Steady-Glo® (all from Promega), and Steadylite plus™ luminescence reporter gene system (Perkin Elmer). Assays → TNF-α neutralization (cytotoxicity, apoptosis, reporter gene): measuring the extent of residual TNF-α-induced cytotoxicity and apoptosis in the presence of anti-TNF-α biologics; TNF-α-induced ADCC: measuring the extent of effector cell cytotoxicity on htmTNF-α expressing target cells, in the presence of anti-TNF-biologics; TNF-α-induced CDC: measuring the extent of cytotoxicity in the presence of human serum as a source of complement and anti-TNF biologics.

Table 1.
Most frequently used routine cell-based bioassays for assessing the TNF neutralization potency (A), ADCC (B), and CDC (C) of anti-TNF biologics and their biosimilars [14, 15].

of collagen and proteoglycans [24]. *In vitro*-cultured chondrocytes have already provided useful models to study their response to microenvironment alterations [25]. However, we have extended their *in vitro* use to efficacy testing of anti-TNF-α drugs [26–28]. First, we have established a 2D *in vitro* model by culturing human primary chondrocytes in monolayer cultures and later upgraded it to a 3D cell model, which better mimics the organization of these cells in native cartilage. For this purpose, we chose a combination of physiologically relevant cell sources and a gene expression assessment technique (qRT-PCR), which enables analyses of up- or downregulated genes in comparison to measurable changes in secreted proteins or cell numbers [29]. We have selected and screened 42 genes involved in immune responses, extracellular matrix remodeling, stress response, signaling pathways, expression of adhesion and other molecules, responding to a pathogenic inflammatory environment that was artificially created with the addition of rhTNF-α.

For the establishment of our 2D model, two types of cells were used. Normal, healthy chondrocytes (NCs) were obtained from surplus cartilage biopsies of patients scheduled for an autologous chondrocyte implantation procedure or were acquired postmortem from donors with healthy cartilage, in accordance with National Medical Ethics Committee approvals. On the other hand, osteoarthritic chondrocytes (OACs) were obtained from cartilage samples of patients undergoing total knee replacement surgery, in accordance with National Medical Ethics Committee approval. Following chondrocyte isolation and cultivation, confluent cell cultures were incubated in serum-free conditions with 1 ng/mL of rhTNF-α (PeproTech, USA) ± 1 μg/mL of each of the two anti-TNF-α biologicals tested, infliximab (IFX; Remicade®, Centocor, Netherlands) and etanercept (ETA; Enbrel®, Wyeth Pharmaceuticals, UK). After 24 h of incubation, chondrocytes and cell culture media were sampled for gene and protein expression analyses, respectively. In experiments using OACs, only the most relevant genes were selected and analyzed. Names and symbols of screened genes are presented in **Table 2**. Data were analyzed by applying the $2^{-\Delta\Delta Cq}$ formula (ABI PRISM® 7700 Sequence Detection System User Bulletin #2) with the nontreated chondrocyte samples used for normalization. Results are presented as relative quantities (RQ) or Log2 relative quantity values (Log2 RQ). For protein expression analysis, a custom antibody array (RayBiotech, USA) was designed to detect interleukin-1 receptor antagonist (IL-1Ra), interleukin-1β (IL-1β), interleukin-6 (IL-6), interleukin-8 (IL-8), matrix metalloproteinase-1 (MMP-1), matrix metalloproteinase-3 (MMP-3), matrix metalloproteinase-13 (MMP-13), monocyte chemoattractant protein-1 (MCP-1), tissue inhibitor of metalloproteinase-2 (TIMP-2), and vascular cell adhesion protein 1 (VCAM-1). All protein analysis data were normalized to nontreated controls.

The results of the first set of experiments obtained after stimulation of cultured NCs and OACs with rhTNF-α and after their preincubation with a combination of rhTNF-α and IFX or ETA are presented in **Figure 2** (graphs A and B, respectively). Upon TNF-α stimulation of NCs, the highest gene upregulation was observed for *IL8* and *MMP1* with a >1000-fold change. A very high upregulation (≥80-fold change) was also observed for *IL6*, *IL32*, *MMP3*, *MMP13*, *TLR2*, and *MCP1* genes (**Figure 2**, graph Aa). We considered the differences between treated and nontreated cells as biologically significant whenever the calculated fold change was ≥2, which equals a Log2-fold change of ≥1 unit on a logarithmic scale. Next, we examined the neutralization efficacy of IFX and ETA by monitoring a decrease in TNF-α-induced gene expressions. Although IFX reduced the expression of TNF-α-upregulated genes, some of them remained more expressed when compared to nontreated cell samples (**Figure 2**, graph Ab). On the other hand, ETA completely abolished the TNF-α-mediated up- and downregulation of the tested genes (**Figure 2**, graph Ac). Altogether, our results revealed differential sTNF-α

Gene Symbol	Gene Name	Gene Symbol	Gene Name
ACAN	aggrecan	MMP1	matrix metallopeptidase 1 (interstitial collagenase)
ADAM17	ADAM metallopeptidase domain 17	MMP10	matrix metallopeptidase 10 (stromelysin 2)
ADAMTS4	ADAM metallopeptidase with thrombospondin type 1 motif, 4	MMP13	matrix metallopeptidase 13 (collagenase 3)
ADAMTS5	ADAM metallopeptidase with thrombospondin type 1 motif, 5	MMP3	matrix metallopeptidase 3 (stromelysin 1, progelatinase)
BGN	biglycan	NFKB1	nuclear factor of kappa light polypeptide gene enhancer in B-cells 1
CCL13	chemokine (C-C motif) ligand 13	NOS2	nitric oxide synthase 2, inducible
CD44	CD44 molecule (Indian blood group)	PTGS2	prostaglandin-endoperoxide synthase 2 (prostaglandin G/H synthase and cyclooxygenase)
CHUK	conserved helix-loop-helix ubiquitous kinase	RUNX2	runt-related transcription factor 2
COL11A2	collagen, type XI, alpha 2	SOX9	SRY (sex determining region Y)-box 9
COL12A1	collagen, type XII, alpha 1	TIMP1	TIMP metallopeptidase inhibitor 1
COL9A3	collagen, type IX, alpha 3	TIMP2	TIMP metallopeptidase inhibitor 2
COMP	cartilage oligomeric matrix protein	TLR1	toll-like receptor 1
DCN	decorin	TLR2	toll-like receptor 2
HAPLN1	hyaluronan and proteoglycan link protein 1	TLR4	toll-like receptor 4
IL1B	interleukin 1, beta	TNF	tumor necrosis factor
IL1RA	interleukin 1 receptor antagonist	TNFRSF1A	tumor necrosis factor receptor superfamily, member 1A
IL32	interleukin 32	TNFRSF1B	tumor necrosis factor receptor superfamily, member 1B
IL6	interleukin 6 (interferon, beta 2)	VCAM1	vascular cell adhesion molecule 1
IL8	interleukin 8	VEGFA	vascular endothelial growth factor A
MAP2K3	mitogen-activated protein kinase kinase 3		
MAPK1	mitogen-activated protein kinase 1		
MMP1	matrix metallopeptidase 1 (interstitial collagenase)		
MCP1	chemokine (C-C motif) ligand 2		

Table 2.
List of genes analyzed in qRT-PCR experiments.

neutralizing potency of IFX and ETA at the level of gene expression patterns. The observed changes in gene expression were then also confirmed with a protein expression assay.

Because NCs are difficult to obtain, we performed the same IFX and ETA neutralization experiments with rhTNF-α-treated OACs, however, to a lesser extent. A selected group of the most responsive genes were tested using OAC biological samples from four donors (**Figure 2B**). We observed a similar response to NCs when OACs were treated by rhTNF-α alone (**Figure 2**, graph Ba) and after their preincubation with a combination of rhTNF-α and IFX or ETA (**Figure 2**, graphs Bb and Bd, respectively). In **Figure 2**, graphs Bc and Be show the responses of OACs after their exposure to each individual biological drug. The cartilage of OA patients represents a biological waste material, which can be obtained in joint replacement surgeries. Despite changes in gene expression observed during the *in vitro* cultivation of OACs, the stimulation with rhTNF-α reverted them back to their inflammatory phenotype. Our results, employing anti-TNF-α biologics,

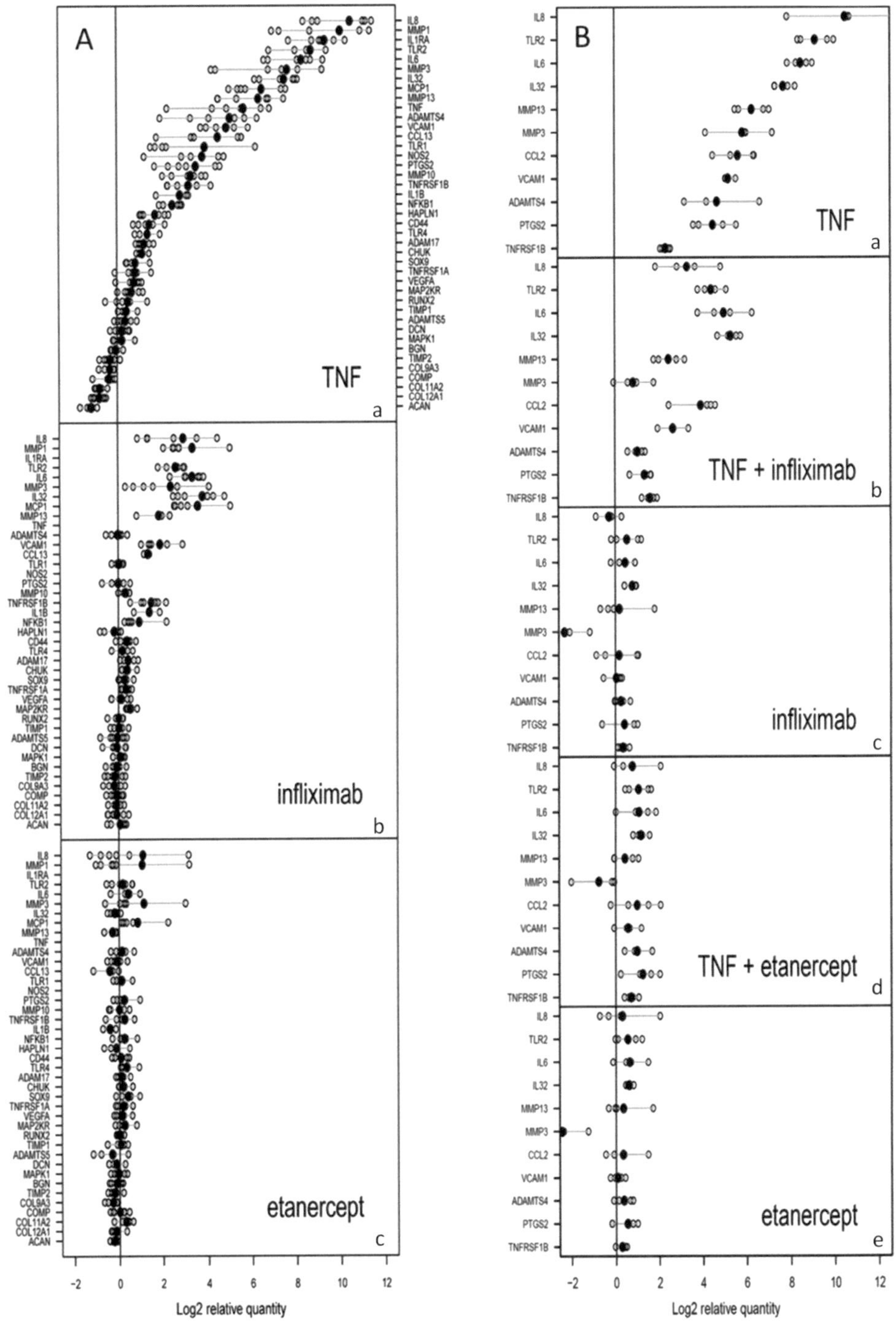

Figure 2.
Exposure of in vitro*-cultured normal healthy chondrocytes (NCs) (A) and osteoarthritic chondrocytes (OACs) (B) to rhTNF-α markedly increased the expression of various genes coding for interleukins, matrix metalloproteinases, and other factors involved in inflammation and stress response (graphs Aa and Ba). TNF-α neutralizing effect of IFX (graphs Ab and Bb) and ETA (graphs Ac and Bd) is reflected in significantly decreased gene expressions. Graphs Bc and Be show the responses of OACs following their exposure to each individual biological drug tested. All gene expressions were calibrated to nontreated cells. Log2 RQ values of individual biological samples (○) and their corresponding geometrical means (●) are shown. Original figures used with authors' permission under the terms and conditions of the Creative Commons Attribution (CC BY) license (http://creativecommons.org/licenses/by/4.0/) [27, 28].*

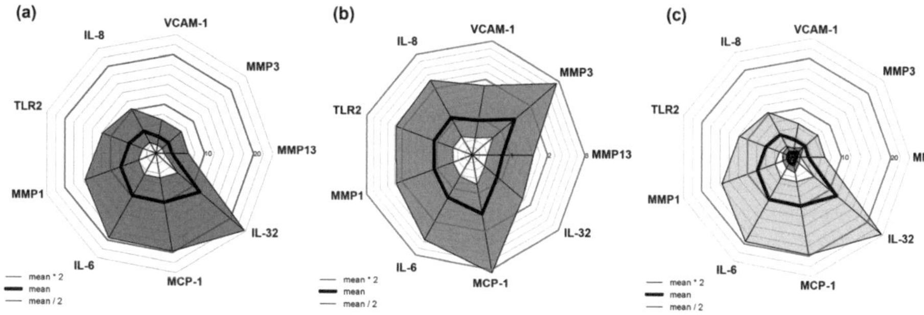

Figure 3.
Graphical representation of results obtained with the statistical model for evaluation of sTNF-α neutralization efficacy of biological drugs. Radar graphs for IFX (a), ETA (b), and both of them (IFX and ETA) (c) present geometrical means of RQ values obtained with six NC biological samples. Corresponding twofold area changes are presented. The overlap of twofold rims is only seen in two (MMP3 and VCAM-1) out of nine genes, indicating higher sTNF-α neutralization efficacy of ETA. Value 0 in the center of the radar graphs denotes total inhibition of gene expression. Please note the difference in scales: (b) intervals of 0.2 units, (a) and (c) intervals of 2 units. Original figure used with authors' permission under the terms and conditions of the Creative Commons Attribution (CC BY) license (http://creativecommons.org/licenses/by/4.0/) [28].

confirmed that OACs and NCs can be interchangeably used for obtaining valuable preliminary information regarding the neutralization efficacy of these drugs [27].

With the data obtained, we were able to establish a statistical model for the evaluation of IFX and ETA TNF-α neutralization efficacy. Expressions of the nine most representative genes were chosen for a graphical presentation of results. Geometrical means of RQ values were plotted on radial axes of radar graphs and connected by a polygon, forming a distinctive shape. A comparison of shapes obtained with IFX and ETA revealed differences in their inhibition of gene expressions. Value 0, depicted in the center of graphs, represents total gene inhibition. For easier comparisons of results, shaded areas of twofold changes were plotted as well. Arbitrary fold-change cutoffs >2 (0.5 for down- and 2 for upregulated genes) were considered biologically significant. In our experimental conditions, the twofold change rims only overlapped in case of *VCAM1* and *MMP3* gene expressions, indicating that both IFX and ETA inhibit these two genes to a similar extent. However, in the case of *MMP13*, *IL32*, *MCP1*, *IL6*, *MMP1*, *TLR2*, and *IL8*, the inhibition efficacy of ETA was significantly more pronounced. Altogether, in our 2D NC-based model, ETA exhibited higher sTNF-α neutralization efficacy than IFX (**Figure 3**).

The presented statistical model is also suitable for a comparative neutralization efficacy determination of new bioactive molecules and biosimilars relative to well-established and approved biologics, according to effective criteria for the assessment of biosimilarity, nonsimilarity, and incomparability.

5.2 Establishment of a 3D human osteoarthritic model for *in vitro* efficacy testing of anti-TNF-α biologicals, using primary human osteoarthritic chondrocytes and mesenchymal stem cells

As discussed in the introduction, 2D and 3D cell culture conditions have different impacts on cell phenotype and biological behavior, which were also confirmed for primary chondrocytes and chondrogenically differentiated MSCs [30–35]. In the last decade, cell-based research shifted toward 3D tissue/organ models, providing more physiologically realistic biochemical and biomechanical microenvironments. However, besides their biological relevance, in order to meet the expectations of the pharmaceutical industry, drug screening assays should be high-throughput, widely applicable, and low cost. With this in mind, we established a new *in vitro* 3D chondrogenic tissue model which, combined with the qRT-PCR readout method, can be used for preclinical or patient-specific potency assessment of anti-TNF-α and anti-interleukin-1β biological drugs (anti-IL-1β) [26]. For establishing this model, we used human OACs and chondrogenically differentiated MSCs.

As already stated, OACs represent an attractive source of cells for cell-based models as besides being rather easily accessible and free of ethical concerns, they are also genetically stable during their long-term *in vitro* expansion [36, 37]. Reports show that MSCs isolated from bone marrow of OA patients are capable of producing hyaline cartilage suitable for tissue repair. MSCs obtained from OA and RA patients possess similar chondrogenic potential as those from healthy individuals [38–41]. Therefore, we used paired samples of MSCs and OACs from two donors and a set of genetically mismatched biological samples of patient's OACs and commercially available MSCs. The paired cell sampling approach allowed us to reduce the high patient-to-patient variability, which influences the chondrogenic potential of both OACs and MSCs [42].

Among the numerous commercially available 3D cell culture systems, we have chosen Perfecta 3D® scaffolds (3D Biomatrix Inc., USA) to create tissues in hanging drops. Generation of scaffold-free spheroids of micrometric dimensions (microspheroids) by gravity-enforced self-assembly in hanging drops allows cell aggregation and tissue formation in a natural manner, without interference from the scaffold material [19, 32]. This technique has important advantages, especially the drop size control and consequent uniformity of formed microspheroids. Moreover, it is compatible with automated liquid handling systems, a prerequisite for high-throughput screening in drug discovery. The microspheroid formation in hanging drops mimics the condensation process of MSCs, which is one of the earliest phases of *in vivo* cartilage development [32].

Isolated OACs were first expanded in 2D monolayer cultures and then, from passage 2 and on, 10,000 cells were transferred into each hanging drop. In this way, the loss of chondrogenic phenotype of OACs in 2D was restored in 3D conditions, as already reported [30, 43]. Similarly as in our previously described 2D primary chondrocyte model, the TNF-α neutralizing efficiencies of ADA (Humira®, Abbott Laboratories, USA), ETA (Enbrel®, Immunex Corp., USA), IFX (Remicade®, Janssen Biotech, USA), and the anti-IL-1β drug anakinra (ANA; Kineret®, Swedish Orphan Biovitrum AB, Sweden) were assessed with both cell types by determining the extent of downregulation of six selected genes (*IL6*, *IL8*, *MCP1*, *MMP1*, *MMP13*, and *VCAM1*) [27, 28]. Gene expression was determined after a 24 h incubation of microspheroids in a medium supplemented with 1 ng/mL of an appropriate inflammatory cytokine (rhTNF-α or rhIL-1β; both from PeproTech, USA) or working macrophage conditioned medium (MCM) solution, combined with 1 μg/mL of each individual biological drug tested (**Figure 4**).

According to our criteria, Log_2 RQ ≥ 1 and ≤ -1, TNF-α significantly upregulated the expression of *IL6*, *IL8*, *MCP1*, *MMP1*, *MMP13*, and *VCAM1* genes in the 3D microspheroidal model as well (**Figure 4a**). The same was true when IL-1β or MCM was added to microspheroids. MCM was obtained from cell cultures of the human monocytic cell line THP-1 (ATCC, USA) and represented a rich source of inflammatory cytokines with 0.05 ng/mL TNF-α and 0.45 ng/mL IL-1β, and numerous other growth factors. In terms of influencing gene expression, IL-1β was the most potent inflammation inducer, followed by MCM and then TNF-α. The inflammatory process triggered by each of these three inducers could always be reversed by ADA, IFX, or ETA, as well as ANA (**Figure 4a**). When inflammation was triggered by TNF-α, all tested anti-TNF-α biologics extraordinarily suppressed the expression of monitored genes, sometimes even reaching their constitutively expressed levels (log2 RQ = 0). Similarly, in the presence of IL-1β, ANA markedly reversed the

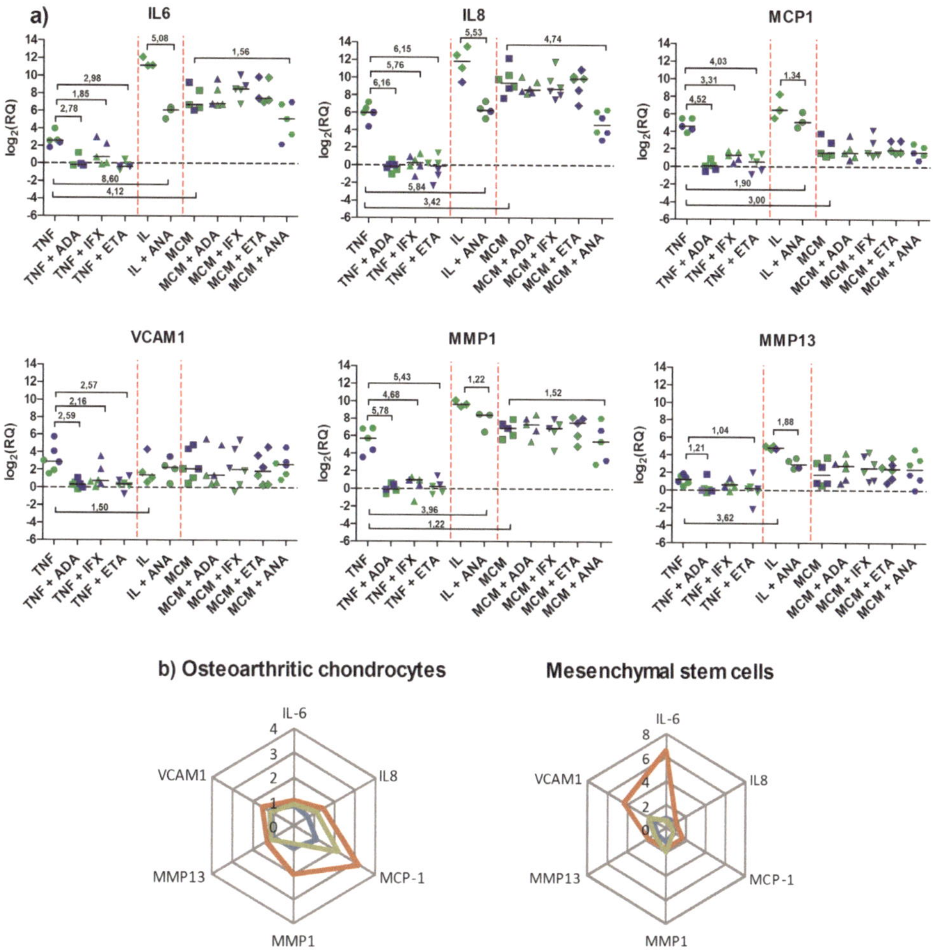

Figure 4.
(a) Gene expression profiles following the addition of inflammatory mediators TNF-α, IL-1β, or MCM working solution and anti-inflammatory biological drugs ADA, IFX, ETA, and ANA. Blue and green dots represent values obtained in microspheroidal chondral tissues made of MSCs and OACs (three donors), respectively. Statistically significant changes, that is, Log2 RQ ≥1 and ≤ −1 are outlined with median values for all groups. (b) Radar graphs representing anti-TNF-α neutralization efficacies of ADA (blue), IFX (red), and ETA (green). Mean RQ values of three biological samples are shown for OAC- and MSC-derived microspheroids. Value 0 in the center of each radar graph represents total inhibition of gene expression. Original figure used with authors' permission under the terms and conditions of the Creative Commons Attribution (CC BY) license (http://creativecommons.org/licenses/by/4.0/) [26].

inflammation process in microspheroids. However, when microspheroids were incubated with MCM, none of the three tested anti-TNF-α drugs were successful in diminishing its inflammatory effect. Conversely, ANA could downregulate the expression of *IL6*, *IL8*, and *MMP1* genes. The described changes at the gene level were accompanied by significant differences in the expression of IL6, IL8, and MCP1 proteins, detected in supernatants of microspheroid cultures, 24 h after their incubation with a given inflammatory agent ± selected anti-inflammatory biological drug [26]. Moreover, these results were additionally supported by the amount of glycosaminoglycans present in chondral spheroids composed of 100,000 OACs treated with various combinations of a particular inflammatory agent ± a given anti-inflammatory biologic, for a period of 3 weeks.

When microspheroids were incubated with MCM, a superior anti-IL-1β neutralization capacity of ANA compared to the three tested anti-TNF-α biologics was observed. This difference was probably due to the fact that MCM contained a much higher concentration of IL-1β (0.45 ng/mL) than TNF-α (0.05 ng/mL). Nevertheless, these concentrations of both cytokines are much higher than those measured in synovial fluids of OA and RA patients (0.028 ng/mL TNF-α and 0.1 ng/mL IL-1β) [44]. Although MCM proved to be an excellent *in vitro* inducer of inflammation, its use for potency testing of anti-inflammatory biologicals targeting a specific cytokine is questionable. In fact, from the multiple synergistic proinflammatory effects evoked by different biogenic factors present in MCM, it is very hard to define the potency of a biological targeting a single inflammatory factor.

In our 3D microspheroidal rhTNF-α-induced inflammation model, the neutralization capacity of ADA was superior over that of ETA and the even weaker IFX (**Figure 4b**). Similar results were obtained with both microspheroids, regardless of whether they were made of OACs or chondrogenically differentiated MSCs. The observed differences in neutralizing efficiencies of ADA, ETA, and IFX can be attributed to differences in their molecular structures and sTNF-α-binding affinities [45]. The superior anti-TNF-α efficacy of ADA over ETA and IFX has already been reported together with data, showing that the sTNF-α-binding affinity of ADA is higher for ADA (Kd = 7.05×10^{-11}) than ETA (Kd = 2.35×10^{-11}) and IFX (Kd = 1.17×10^{-10}) [46–48]. However, according to our criterion, a particular biologic would be statistically more efficient than the compared one if it would cause a ≥2-fold decrease in a selected gene expression. This was not the case in any of our 3D microspheroidal model experiments. Consequently, we assumed that the observed differences in TNF-α neutralizing potency of ADA, ETA, and IFX were comparable (**Figure 4b**). Interestingly, although we showed in our 2D OACs model that ETA was significantly more efficient than IFX, the same kind of experiments carried out in a 3D microspheroidal model did not confirm this finding [26, 27]. We assume that compared to the 2D model, the diffusion of tested biologics in our 3D microspheroidal model was much slower and limited. Undoubtedly, the 3D model better resembles *in vivo* conditions and therefore has a higher relevance. Thus, we concluded that 2D cell culture models may be useful for obtaining preliminary data regarding the anti-inflammatory effects of a particular biological drug, while 3D microtissue models enable more relevant insights in drug-tissue interactions and possible outcomes *in vivo*. The results obtained with our 3D microspheroidal model are also supported by the outcomes of clinical studies conducted on patients with RA, where the efficacies of anti-TNF-α biologics proved to be comparable [49].

We found that OACs and chondrogenically differentiated MSCs are suitable sources for hanging drop chondral 3D microspheroid cultures formation, which are useful for the assessment of neutralization potencies of anti-inflammatory biologics [26]. Although the use of these two types of microspheroids resulted in different gene expression profiles following their incubation with tested combinations of rhTNF-α, and each of the three tested anti-TNF-α biological drugs (**Figure 4b**), these differences were rather small. Therefore, we concluded that MSCs can be used as an alternative and probably even more accessible cell source for *in vitro* testing of neutralization potency of anti-TNF-α biologics. The main advantages of our 3D model are the use of small amounts of human cells and cytokines, personalized testing approach, and the possibility of automation. In addition, the presented approach can also be used as a platform for testing other anti-inflammatory biologics with different mechanisms of action, as shown for ANA, the antagonist of IL-1β.

6. Conclusion

Cell-based assays are complex analytical tools, susceptible to multiple variables that are virtually impossible to control. Therefore, they have to be precise, reliable, and well standardized so that the results are reproducible and can be compared among different laboratories. When used for drug potency testing, such assays usually rely on the use of reference standards. Recently, the WHO has prepared two international standards for the two anti-TNF-α biologics, etanercept and infliximab. These have been tested by several laboratories within an international collaborative study using a number of different cell-based assays [12, 13]. In this chapter, we have presented an overview of the most routinely used tests for potency testing of anti-TNF-α biologics, which measure *in vitro* responses of nonmanipulated or genetically engineered human and animal cell lines, with various readout systems.

Nowadays, with an expanding personal medicine approach, laboratory assay-guided pharmacotherapeutical strategies are becoming more and more important. In order to obtain relevant data on drug potencies for a particular patient, these kinds of tests should be based on the patient's own, that is, autologous primary cells, as these can significantly reduce costs and enable safer and more effective therapies. Therefore, we dedicated a part of this chapter to our experience in establishing *in vitro* 2D monolayer cultures consisting of normal and OA chondrocytes and 3D microspheroidal chondral tissues, formed from OACs or chondrogenically differentiated bone-marrow-derived MSCs, and their use for testing anti-TNF-α efficacy of adalimumab, etanercept, and infliximab. The qRT-PCR technique was applied as a readout system for assessing differences in selected gene expressions. The obtained data led us to the establishment of an original statistical method, which was used for the evaluation and comparison of results.

Acknowledgements

The authors wish to express their gratitude to Prof. Andrej Blejec, Dr. Miomir Knežević, Asst. Prof. Nevenka Kregar Velikonja, and Prof. Gordana Vunjak-Novakovic for their contribution to our common, published original research work, partly presented in this chapter. We are also grateful to Rosmarijn E. Vandenbroucke for approving the use of a picture presented in **Figure 1** and to Andrej Branc for proofreading the manuscript.

Conflict of interest

The authors declared that no competing interests exist.

Appendices and nomenclature

2D	two-dimensional
3D	three-dimensional
ADA	adalimumab
ANA	anakinra
ETA	etanercept
IFX	infliximab
IL-1β	interleukin 1β
MCM	macrophage conditioned medium
MSCs	mesenchymal stem cells
NCs	normal human articular chondrocytes
OACs	osteoarthritic human articular chondrocytes
RQ	relative quantity of gene expression

rhTNF-α	recombinant human tumor necrosis factor α
sTNF-α	soluble form of tumor necrosis factor α
tmTNF-α	transmembrane form of tumor necrosis factor α

Author details

Sara Žigon-Branc[1]†, Ariana Barlič[2]† and Matjaž Jeras[3]*

1 Institute of Materials Science and Technology, Technische Universität Wien, Vienna, Austria

2 Educell Cell Therapy Service Ltd., Trzin, Slovenia

3 Faculty of Pharmacy, University of Ljubljana, Ljubljana, Slovenia

*Address all correspondence to: matjaz.jeras@ffa.uni-lj.si

† These authors contributed equally to this work

References

[1] Aggarwal BB, Gupta SC, Kim JH. Historical perspectives on tumor necrosis factor and its superfamily: 25 years later, a golden journey. Blood. 2012;**119**:651-665. DOI: 10.1182/blood-2011-04-325225

[2] Steeland S, Libert C, Vandenbroucke RE. A new venue of TNF targeting. International Journal of Molecular Sciences. 2018;**19**:1442. DOI: 10.3390/ijms19051442

[3] Yan L, Hu R, Tu S, Cheng W-J, Zheng Q, Wang J-W, et al. Establishment of a cell model for screening antibody drugs against rheumatoid arthritis with ADCC and CDC. International Journal of Clinical and Experimental Medicine. 2015;**8**:20065-20071

[4] Mitoma H, Horiuchi T, Tsukamoto H, Ueda N. Molecular mechanisms of action of anti-TNF-α agents—Comparison among therapeutic TNF-α antagonists. Cytokine. 2018;**101**:56-63. DOI: 10.1016/j.cyto.2016.08.014

[5] Chen X, Oppenheim JJ. Targeting TNFR2, an immune checkpoint stimulator and oncoprotein, is a promising treatment for cancer. Science Signaling. 2017;**10**:eaal2328. DOI: 10.1126/scisignal.aal2328

[6] Torrey H, Butterworth J, Mera T, Okubo Y, Wang L, Baum D, et al. Targeting TNFR2 with antagonistic antibodies inhibits proliferation of ovarian cancer cells and tumor-associated T_{regs}. Science Signaling. 2017;**10**:eaaf8608. DOI: 10.1126/scisignal.aaf8608

[7] Vos ACW, Wildenberg ME, Duijvestein M, Verhaar AP, van den Brink GR, Hommes DW. Anti-tumor necrosis factor-α antibodies induce regulatory macrophages in an Fc region-dependent manner. Gastroenterology. 2011;**140**:221-230.e3. DOI: 10.1053/j.gastro.2010.10.008

[8] Nguyen DX, Ehrenstein MR. Anti-TNF drives regulatory T cell expansion by paradoxically promoting membrane TNF–TNF-RII binding in rheumatoid arthritis. The Journal of Experimental Medicine. 2016;**213**:1241-1253. DOI: 10.1084/jem.20151255

[9] Nadkarni S, Mauri C, Ehrenstein MR. Anti-TNF-α therapy induces a distinct regulatory T cell population in patients with rheumatoid arthritis via TGF-β. The Journal of Experimental Medicine. 2007;**204**:33-39. DOI: 10.1084/jem.20061531

[10] Shealy DJ, Cai A, Staquet K, Baker A, Lacy ER, Johns L, et al. Characterization of golimumab, a human monoclonal antibody specific for human tumor necrosis factor α. MAbs. 2010;**2**:428-439

[11] Paleolog E. Target effector role of vascular endothelium in the inflammatory response: Insights from the clinical trial of anti-TNF alpha antibody in rheumatoid arthritis. Molecular Pathology. 1997;**50**:225-233

[12] Wadhwa M, Bird C, Dilger P, Rigsby P, Jia H, Gross MEB. Establishment of the first WHO International Standard for etanercept, a TNF receptor II Fc fusion protein: Report of an international collaborative study. Journal of Immunological Methods. 2017;**447**:14-22. DOI: 10.1016/j.jim.2017.03.007

[13] Metcalfe C, Dougall T, Bird C, Rigsby P, Behr-Gross M-E, Wadhwa M, et al. The first World Health Organization International Standard for infliximab products: A step towards maintaining harmonized biological activity. MAbs. 2019;**11**:13-25. DOI: 10.1080/19420862.2018.1532766

[14] Hofmann H-P, Kronthaler U, Fritsch C, Grau R, Müller SO, Mayer R, et al. Characterization and non-clinical assessment of the proposed etanercept biosimilar GP2015 with originator etanercept (Enbrel(®)). Expert Opinion on Biological Therapy. 2016;**16**: 1185-1195. DOI: 10.1080/14712598.2016.1217329

[15] Lallemand C, Kavrochorianou N, Steenholdt C, Bendtzen K, Ainsworth MA, Meritet J-F, et al. Reporter gene assay for the quantification of the activity and neutralizing antibody response to TNFα antagonists. Journal of Immunological Methods. 2011;**373**: 229-239. DOI: 10.1016/j.jim.2011.08.022

[16] Mitoma H, Horiuchi T, Tsukamoto H, Tamimoto Y, Kimoto Y, Uchino A, et al. Mechanisms for cytotoxic effects of anti-tumor necrosis factor agents on transmembrane tumor necrosis factor α-expressing cells: Comparison among infliximab, etanercept, and adalimumab. Arthritis and Rheumatism. 2008;**58**:1248-1257. DOI: 10.1002/art.23447

[17] Ueda N, Tsukamoto H, Mitoma H, Ayano M, Tanaka A, Ohta S, et al. The cytotoxic effects of certolizumab pegol and golimumab mediated by transmembrane tumor necrosis factor α. Inflammatory Bowel Diseases. 2013;**19**: 1224-1231. DOI: 10.1097/MIB.0b013e318280b169

[18] Moore M, Ferguson J, Burns C. Applications of cell-based bioassays measuring the induced expression of endogenous genes. Bioanalysis. 2014;**6**: 1563-1574. DOI: 10.4155/bio.14.98

[19] Rimann M, Graf-Hausner U. Synthetic 3D multicellular systems for drug development. Current Opinion in Biotechnology. 2012;**23**:803-809. DOI: 10.1016/j.copbio.2012.01.011

[20] Baker BM, Chen CS. Deconstructing the third dimension—How 3D culture microenvironments alter cellular cues. Journal of Cell Science. 2012;**125**: 3015-3024. DOI: 10.1242/jcs.079509

[21] Edmondson R, Broglie JJ, Adcock AF, Yang L. Three-dimensional cell culture systems and their applications in drug discovery and cell-based biosensors. Assay and Drug Development Technologies. 2014;**12**: 207-218. DOI: 10.1089/adt.2014.573

[22] Duval K, Grover H, Han L-H, Mou Y, Pegoraro AF, Fredberg J, et al. Modeling physiological events in 2D vs. 3D cell culture. Physiology. 2017;**32**:266-277. DOI: 10.1152/physiol.00036.2016

[23] Alépée N. State-of-the-art of 3D cultures (organs-on-a-chip) in safety testing and pathophysiology. ALTEX. 2014;**31**(4):441-477. DOI: 10.14573/altex1406111

[24] Sophia Fox AJ, Bedi A, Rodeo SA. The basic science of articular cartilage: Structure, composition, and function. Sports Health: Multidisciplinary Approach. 2009;**1**:461-468. DOI: 10.1177/1941738109350438

[25] Otero M, Favero M, Dragomir C, Hachem KE, Hashimoto K, Plumb DA, et al. Human chondrocyte cultures as models of cartilage-specific gene regulation. In: Mitry RR, Hughes RD, editors. Human Cell Culture Protocols. Vol. 806. Totowa, NJ: Humana Press; 2012. pp. 301-336

[26] Žigon-Branc S, Barlič A, Knežević M, Jeras M, Vunjak-Novakovic G. Testing the potency of anti-TNF-α and anti-IL-1β drugs using spheroid cultures of human osteoarthritic chondrocytes and donor-matched chondrogenically differentiated mesenchymal stem cells. Biotechnology Progress. 2018;**34**: 1045-1058. DOI: 10.1002/btpr.2629

[27] Žigon-Branc S, Jeras M, Blejec A, Barlič A. Applicability of human osteoarthritic chondrocytes for in vitro

efficacy testing of anti-TNFα drugs. Biologicals. 2017;**45**:96-101. DOI: 10.1016/j.biologicals.2016.09.013

[28] Barlič A, Žigon S, Blejec A, Kregar Velikonja N. Gene expression of cultured human chondrocytes as a model for assessing neutralization efficacy of soluble TNFα by TNFα antagonists. Biologicals. 2015;**43**: 171-180. DOI: 10.1016/j.biologicals.2015.03.001

[29] Burns CJ, Silva MMCG, Gray E, Robinson CJ. Quantitative RT-PCR as an alternative to late-stage bioassays for vascular endothelial growth factor. Journal of Pharmaceutical and Biomedical Analysis. 2008;**47**:460-468. DOI: 10.1016/j.jpba.2008.02.011

[30] Caron MMJ, Emans PJ, Coolsen MME, Voss L, Surtel DAM, Cremers A, et al. Redifferentiation of dedifferentiated human articular chondrocytes: Comparison of 2D and 3D cultures. Osteoarthritis and Cartilage. 2012;**20**:1170-1178. DOI: 10.1016/j.joca.2012.06.016

[31] Mata-Miranda MM, Martinez-Martinez CM, Noriega-Gonzalez JE, Paredes-Gonzalez LE, Vázquez-Zapién GJ. Morphological, genetic and phenotypic comparison between human articular chondrocytes and cultured chondrocytes. Histochemistry and Cell Biology. 2016;**146**(2):183-189. DOI: 10.1007/s00418-016-1437-4

[32] Lehmann M, Martin F, Mannigel K, Kaltschmidt K, Sack U, Anderer U. Three-dimensional scaffold-free fusion culture: The way to enhance chondrogenesis of in vitro propagated human articular chondrocytes. European Journal of Histochemistry. 2013;**57**:31. DOI: 10.4081/ejh.2013.e31

[33] Bhumiratana S, Vunjak-Novakovic G. Engineering physiologically stiff and stratified human cartilage by fusing condensed mesenchymal stem cells. Methods. 2015;**84**:109-114. DOI: 10.1016/j.ymeth.2015.03.016

[34] Baraniak PR, McDevitt TC. Scaffold-free culture of mesenchymal stem cell spheroids in suspension preserves multilineage potential. Cell and Tissue Research. 2012;**347**:701-711. DOI: 10.1007/s00441-011-1215-5

[35] Sart S, Tsai A-C, Li Y, Ma T. Three-dimensional aggregates of mesenchymal stem cells: Cellular mechanisms, biological properties, and applications. Tissue Engineering. Part B, Reviews. 2014;**20**:365-380. DOI: 10.1089/ten.teb.2013.0537

[36] Laganà M, Arrigoni C, Lopa S, Sansone V, Zagra L, Moretti M, et al. Characterization of articular chondrocytes isolated from 211 osteoarthritic patients. Cell and Tissue Banking. 2014;**15**:59-66. DOI: 10.1007/s10561-013-9371-3

[37] Neri S, Mariani E, Cattini L, Facchini A. Long-term in vitro expansion of osteoarthritic human articular chondrocytes do not alter genetic stability: A microsatellite instability analysis. Journal of Cellular Physiology. 2011;**226**:2579-2585. DOI: 10.1002/jcp.22603

[38] Tallheden T, Bengtsson C, Brantsing C, Sjogren-Jansson E, Carlsson L, Peterson L, et al. Proliferation and differentiation potential of chondrocytes from osteoarthritic patients. Arthritis Research & Therapy. 2005;7:R560-R568

[39] Kafienah W, Mistry S, Dickinson SC, Sims TJ, Learmonth I, Hollander AP. Three-dimensional cartilage tissue engineering using adult stem cells from osteoarthritis patients. Arthritis and Rheumatism. 2007;**56**:177-187. DOI: 10.1002/art.22285

[40] Jagielski M, Wolf J, Marzahn U, Völker A, Lemke M, Meier C, et al. The

influence of IL-10 and TNFα on chondrogenesis of human mesenchymal stromal cells in three-dimensional cultures. International Journal of Molecular Sciences. 2014;**15**: 15821-15844. DOI: 10.3390/ijms150915821

[41] Dudics V, Kunstár A, Kovács J, Lakatos T, Géher P, Gömör B, et al. Chondrogenic potential of mesenchymal stem cells from patients with rheumatoid arthritis and osteoarthritis: Measurements in a microculture system. Cells, Tissues, Organs. 2009;**189**:307-316. DOI: 10.1159/000140679

[42] Agar G, Blumenstein S, Bar-Ziv Y, Kardosh R, Schrift-Tzadok M, Gal-Levy R, et al. The chondrogenic potential of mesenchymal cells and chondrocytes from osteoarthritic subjects: A comparative analysis. Cartilage. 2011;**2**: 40-49. DOI: 10.1177/1947603510380899

[43] Schulze-Tanzil G, de Souza P, Villegas Castrejon H, John T, Merker H-J, Scheid A, et al. Redifferentiation of dedifferentiated human chondrocytes in high-density cultures. Cell and Tissue Research. 2002;**308**:371-379. DOI: 10.1007/s00441-002-0562-7

[44] Westacott CI, Whicher JT, Barnes IC, Thompson D, Swan AJ, Dieppe PA. Synovial fluid concentration of five different cytokines in rheumatic diseases. Annals of the Rheumatic Diseases. 1990;**49**:676-681

[45] Tracey D, Klareskog L, Sasso EH, Salfeld JG, Tak PP. Tumor necrosis factor antagonist mechanisms of action: A comprehensive review. Pharmacology & Therapeutics. 2008;**117**:244-279. DOI: 10.1016/j.pharmthera.2007.10.001

[46] Hu S, Liang S, Guo H, Zhang D, Li H, Wang X, et al. Comparison of the inhibition mechanisms of adalimumab and infliximab in treating tumor necrosis factor α-associated diseases from a molecular view. The Journal of Biological Chemistry. 2013;**288**: 27059-27067. DOI: 10.1074/jbc.M113.491530

[47] Scallon B, Cai A, Solowski N, Rosenberg A, Song X-Y, Shealy D, et al. Binding and functional comparisons of two types of tumor necrosis factor antagonists. The Journal of Pharmacology and Experimental Therapeutics. 2002;**301**:418-426

[48] Granneman RG, Zhang YM, Noertersheuser PA, Velagapudi RB, Awni WM, Locke CS, et al. Pharmacokinetic/pharmacodynamic (PK/PD) relationships of adalimumab (HUMIRA™, Abbott) in rheumatoid arthritis (RA) patients during phase II/III clinical trials. In: Arthritis and Rheumatism. Vol. 48. Div John Wiley & Sons Inc, 605 Third Ave, New York, NY 10158–0012 USA: Wiley-Liss; 2003. pp. S140-S141

[49] Hyrich KL, Lunt M, Watson KD, Symmons DPM, Silman AJ. British Society for Rheumatology Biologics Register. Outcomes after switching from one anti-tumor necrosis factor alpha agent to a second anti-tumor necrosis factor alpha agent in patients with rheumatoid arthritis: Results from a large UK national cohort study. Arthritis and Rheumatism. 2007;**56**:13-20. DOI: 10.1002/art.22331

Chapter 3

Tumor Necrosis Factor Alpha: A Major Cytokine of Brain Neuroinflammation

Mubarak Muhammad

Abstract

Tumor necrosis factor (TNF) is one of the most extensively studied cytokine with about 19 distinct superfamily members and many more to be found. Prominent among these members is tumor necrosis factor alpha (TNF-α) that is known to be a potent promoter of inflammation, as well as many normal physiological functions in homeostasis and health and antimicrobial immunity. Nuclear factor kappa-light-chain enhancer of activated B cells (NFκB) is one of the most important transcription factors that activate transcription of many proinflammatory genes, and the unraveling of TNF-α induced NFκB activation forms the foundation of TNF-α as major cytokine of neuroinflammation. This review discusses summary of literature on unique role of TNF-α in neuroinflammation and various agents that mediate neuroinflammation via TNF-α modulation.

Keywords: tumor necrosis factor, tumor necrosis factor alpha, neuroinflammation, cytokine, brain, inflammation

1. Introduction

Tumor necrosis factor (TNF) alpha is one of the first discovered cytokines shown by Carswell [1] in 1975 and was named for tumor regression activity induced in the serum of mice treated with *Serratia marcescens* polysaccharide [2]. Cytokines are low-molecular-weight peptides secreted by activated immune cells as well as stromal cells and exerting biological activities through binding to cognate receptors on cell surface. Cytokines are produced by a number of cell types, predominantly leukocytes that regulate a number of physiological and pathological functions including innate immunity, acquired immunity, and a plethora of inflammatory responses [3]. Cytokines excite or hinder the generation, propagation, and differentiation of different associated target cells positive on antigen induction, thus leading to mediation in the activity of diverse other cells involved in the immune response especially the more pronounced macrophages, mast cells, B cells, T cells, and natural killer (NK) cells. Thus, cytokine is regarded as secreted proteins with growth, differentiation, and activation functions that regulate and determine the nature of immune responses [4]. The broad classification of cytokines are termed in a group as follows: interleukin (IL), interferon (IFN), tumor necrosis factor (TNF), colony stimulating factor (CSF), and chemokine and growth factor (GF), and these exerts biological functions through action mode and characteristics as paracrine, autocrine, and endocrine. TNF being one of the prominent cytokine has about 19 different members of the TNF superfamily that includes tumor necrosis factor alpha (TNF-α), tumor necrosis factor beta (TNF-β), TNF-related weak inducer of apoptosis (TWEAK), TNF-related apoptosis-inducing ligand (TRAIL), lymphotoxin-β (LT- β), CD40L, CD30L, 4-1BBL, CD27L, glucocorticoid-induced TND receptor ligand (GITRL), fibroblast-associated ligand (FasL), OX40 ligand (OX40L), LIGHT, A proliferation-inducing ligand (APRIL), B-cell-activating factor (BAFF), receptor activator of NFκB ligand (RANKL), vascular endothelial cell-growth inhibitor (VEGI), and ectodysplasin A ((EDA)–A1, EDA-A2) [2].

TNF-α is a potent mediator of inflammation, as well as many normal physiological functions in homeostasis and health and antimicrobial immunity [5]. Inflammation is a classical host defense response of vascularized living tissue to infection and injury, and in the central nervous system (CNS), the term neuroinflammation is used to denote cellular and inflammatory responses of vascularized neuronal tissue through activation of resident cells in the brain (microglia, astrocytes, and endothelial cells), the recruitment of blood-derived leukocytes including neutrophils, lymphocytes, and macrophages, and a plethora of humoral factors [6, 7]. More appropriately, neuroinflammation is a term used to denote inflammation associated with the brain and is characterized by the activation of microglia and expression of major inflammatory mediators without typical features of peripheral inflammation such as edema and neutrophil infiltration [8]. Neuroinflammation in the brain supposedly has a positive effect such as increasing blood flow and removal of damaged tissue by phagocytosis, but in a disease state, the resulting inappropriate inflammation caused negative effects which by far out weight the positive effect [6].

Nuclear factor kappa-light-chain enhancer of activated B cells (NFκB) otherwise called nuclear factor kappa B is a heterodimer and one of the most important transcription factors that activate transcription of many proinflammatory genes. It is well documented that TNF-α induces at least five different types of signals that include activation of NFκB, apoptosis pathways, extracellular signal-regulated kinase (ERK), p38 mitogen-activated protein kinase (p38MAPK), and c-Jun N-terminal kinase (JNK) [2]. These biological functions of TNF-α makes its role in neuroinflammation critically prominent. It is therefore expedient to elucidate and expand the rational basis of TNF-α as major cytokine of neuroinflammation. Hence, this review discusses summary of literature on unique role of TNF-α in neuroinflammation and various agents that mediate neuroinflammation via TNF-α modulation.

2. Neuroinflammation

Microglia being major immune cells involved in defense in the central nervous system, its activation is considered to be the hallmark of neuroinflammation [7, 9]. Activation of microglia cells constitutes the first key acute response in the brain to external aggression such as acute brain ischemia, traumatic brain injury, or microbial pathogen, and this microglial activation is coupled with subsequent activation of blood-borne monocytes/macrophages to yield a full-blown neuroinflammatory thick rim around ischemia infarct that becomes observable after 1 week in both human and animal models [10]. Microglia in the CNS constitutes 5–15% of total brain population; having share common precursor with peripheral macrophages, they produced transient inflammatory changes like macrophages such as phagocytosis, inflammatory cytokine production, and antigen presentation, normally returning to their basal state when the activation stimulus is resolved [11]. In a disease state such as in the onset of focal cerebral ischemia or traumatic brain injury, however, the microglia response becomes inappropriately more reactive and exaggerated to produce plethora of inflammatory mediators that trigger apoptosis and exaggerate neuronal damage [12]. Therefore, microglia/macrophages are the key immune cells concerned with the protection of brain against injury. Their architectural and functional changes are linked with the liberation of injury signals induced by pathology. These cells are usually responsible for clearance of demised neural cells and allow for restoration of lost neuronal functions. However, when markedly activated by the damage-associated molecular patterns subsequent to a disease state, they can generate a huge amount of proinflammatory cytokines that are capable of interrupting neural cells and the blood-brain barrier, and manipulate neurogenesis [9].

The primary function of microglia in the brain is to control any external aggression and neutralize its effect by a process of phagocytosis, which is a chronologically multistep system including oriented gradient motility (chemotaxis); identification of alien foreign agents by membrane lectins and receptors (recognition); encompassing flow round the injurious foreign agents into a vacuole/phagosome (engulfment); unraveling of intracellular secretor pools (granules); and liberation of innate antibiotics and enzymes into the phagosome, generation of reactive oxygen species by an intricate enzymatic system sequestrated on the phagocyte membrane and/or reactive nitrogen species by an inducible nitric oxide synthase, and decapitating and digestion of engulfed foreign substance in the multifaceted phagolysosomal medium (microbial killing) [13]. Therefore, there are four important events of phagocytosis: chemotaxis, recognition, engulfment, and microbial killing.

Chemotaxis is the immediate restricted, valuable host inflammatory reaction that is initiated by local tenant macrophages, demised cells and tissues, plasma factors, and microbial products. Specifically, the closely generated factors of inflammation (cytokines, activated complement protein, kinins, etc.) and microbial factors construct chemotactic gradients, alter endothelial cell membrane receptors, and encourage decrease of the blood flow. Blood-borne monocytes/macrophages that are rolling along the endothelial surface act in response to the chemotactic and cell-mediated signals and are primarily activated to definitely attach to the endothelium by way of their membrane integrins; the second pace is transendothelial migration, denoted as diapedesis, followed by tilting motility toward the inflammatory site (chemotaxis) [13].

Recognition involves identifying and attachment of particle to be ingested by the microglia/phagocytes. There are two methods of recognition: opsonin/opsonin dependent/receptor mediated and non-opsonin/opsonin independent. Opsonin/opsonin dependent is where microglia/phagocytes recognize pathogens via their membrane receptors for opsonins (e.g., complement factors C3b and iC3b and Fc component of immunoglobulins), which are present on the microbial surface, while non-opsonin/opsonin independent is where microglia/phagocytes recognize pathogens via microbial and phagocyte lectins [13]. Because microglia/phagocytes express high-affinity receptors for opsonin, the term opsonization is used to indicate a process, whereby injurious foreign particle becomes coated with substance, thereby enhancing its recognition by leukocyte and making it more open to phagocytosis. As aforementioned, the injurious foreign agents or microbes are usually opsonized by specific protein substances such as immunoglobulin G (IgG) antibodies, breakdown product of compliment (C3b), and fibrinogen all of which phagocytes express high-affinity receptors.

Engulfment refers to microglia/phagocyte extension of cytoplasm (pseudopods) flow around the injurious foreign agents or microbes after its binding with phagocyte and subsequent pinches off to form vesicles (phagosome) that enclose the injurious foreign agents or microbes. Phagocyte extensions (pseudopods) finally engulf the injurious foreign agent or microbe in a vacuole and trigger the activation of two functions: the release of granule contents into the phagosome and the oxidative burst. Coiling engulfment is the most frequent unusual uptake: unilateral pseudopods wrap around the microorganism in multiple turns, giving rise to largely self-apposed pseudopodial surfaces [13].

Microbial killing can be achieved through oxygen-dependent or oxygen-independent/non-oxygen-dependent method of pathogen or injurious agent killing. Oxygen dependent involves the use free radicals. A free radical is clearly referred to as atom or molecule having one or more unpaired electrons in valence shell or outer orbit and is competent for autonomous survival [14]. The strange quantities of odd electron(s) possess by a free radical make it unbalanced, short lived, and extremely reactive. This high reactivity makes free radical exert a pull on electrons from further compounds to reach steadiness. The newly pulled attacked molecule loses its electron and becomes a free radical itself, opening a chain of feedback cascade of reaction. Free radicals/oxidants derived from both endogenous sources and exogenous sources have gained importance in the field of biology due to their central role in various physiological conditions as well as their implication in a diverse range of diseases. They include reactive oxygen species (ROS) which are hydroxyl radicals ($^{\cdot}OH$), superoxide (O_2^{-}), hydrogen peroxide (H_2O_2), and reactive nitrogen species (RNS) which are nitric oxide (NO) and peroxynitrite ($OONO^{-}$). At reasonable or little concentrations, ROS/RNS encompasses desirable effects and engage in a variety of physiological purposes such as in immune function (i.e., guard in opposition to pathogenic microorganisms), in certain cellular signaling pathways, in mitogenic reaction, and in redox directive. Conversely, at excessively elevated concentrations, both ROS and RNS lead to oxidative stress and nitrosative stress, respectively, that potentially cause adverse effect to biological molecules [14].

The mechanism of oxygen-dependent microbial killing is initiated after engulfment where oxygen burst is activated to cause increase in oxygen consumption (50- to 100-fold increase) and metabolism; this leads to massive production of nicotinamide adenosine diphosphate (NADP) as by-product of adenosine triphosphate (ATP) generation by oxidative phosphorylation. The oxygen burst is unrelated to mitochondrial respiration and reflects the activity of the NADPH oxidase system in the cytosol and membrane constituents, which are separated in resting microglia/phagocytes and are reassembled upon microglia/phagocytes activation. The generated NADP through NADPH oxidase enzyme activity generates superoxide (O_2^{-}) which is further converted to hydrogen peroxide (H_2O_2) either spontaneously or through enzymatic catalysis of superoxide dismutase (SOD) enzyme by combining with hydrogen ion (H^{+}). Both hydrogen peroxide (H_2O_2) and superoxide (O_2^{-}) can cause microbial killing. For instance, H_2O_2 in the presence of myeloperoxidase (MPO) released from microglia/phagocytes azurophilic (primary) granules and a halide generates very potent oxidizing agents such as hypochlorous acid (HOCl) and chloramines [13]. Other oxidative species such as singlet oxygen has been suggested to be important for microbial killing through the formation of ozone [15].

Non-oxygen-dependent/oxygen-independent microbial killing is mediated by protein molecule and other factors that are mostly found within the lysosome such as lysozyme, lactoferrin, and elastase. Lysozyme is an enzyme that hydrolyzes N-acetyl glucosamine bond found in glycopeptide coat of all bacterial cell wall. Thus, non-oxygen-dependent/oxygen-independent microbial killing is dependent on protein and peptide antibiotics such as bactericidal permeability-increasing protein, cationic antimicrobial protein 37, and defensins that are stored in peroxidase-positive (azurophilic, primary) granules where they are together localize with active proteases such as elastase, cathepsin G, and proteinase 3. The synergistic interaction of oxygen-dependent and non-oxygen-dependent/oxygen-independent microbial killing systems generally results in pathogen killing [13].

Pathological consequences that result from a disease state of the brain, however, make microglia response becomes inappropriately exaggerated. Microglia when transformed into phagocytes can release a variety of substances many of which are cytotoxic and/or cytoprotective. While cytoprotective substances include neurotrophic molecules such as brain-derived neurotrophic factor (BDNF), insulin-like growth factor I (IGF-I), several other growth factors, and anti-inflammatory factors, cytotoxic substances include proinflammatory cytokines such as TNF-α, IL-1β, and IL-6 as well as other potential cytotoxic molecules including nitric oxide (NO), reactive oxygen species (ROS), and prostanoids. The uniquely outburst cytokines extensively studied in acute ischemic stroke are tumor necrosis factor-α (TNF-α); the interleukins (IL), IL-1β, IL-6, IL-20, and IL-10; and transforming growth factor (TGF)-β. Although IL-1β and TNF-α are proinflammatory that appears to exacerbate cerebral injury, TGF-β and IL-10 are anti-inflammatory that may exert neuroprotective effects, and IL-6 has both pro- and anti-inflammatory effects [16]. Astrocytes, like microglia, are also capable of secreting inflammatory factors such as cytokines, chemotaxis cytokines (chemokines), and NO in response to brain pathological state.

3. Agents that mediate neuroinflammation via TNF-α modulation

Table 1 reveals researches of various agents that mediate neuroinflammation via TNF modulation.

Treatment	Experimental model	Related TNF finding	References
Puerarin	Stroke model of rat middle cerebral artery occlusion	Modulate neuroinflammation by mark reduction in mRNA expression of tumor necrosis factor-α (TNF-α)	[17]
Edaravone and scutellarin	Stroke model of rat intraluminal middle cerebral artery occlusion	Modulate neuroinflammation by attenuating expression levels of TNF-α	[18]
Matrix metalloproteinase-8 inhibitor (M8I)	Stroke model of rat middle cerebral artery occlusion	Modulate neuroinflammation by abrogating TNF-α expression	[19]
Wogonin (5,7-dihydroxy-8-methoxyflavone)	Stroke model of rat middle cerebral artery occlusion	Modulate neuroinflammation by decrease in production of TNF-α	[20]
Nicotine	Stroke model of rat global cerebral ischemia	Modulate neuroinflammation by significant reduction of enhanced expression of tumor necrosis factor alpha (TNF-α)	[21]
Glycyrrhizin (GRZ)	Brain cognitive impairment and neuroinflammation of lipopolysaccharide treated Mice	Modulate neuroinflammation through inhibition of proinflammatory TNF-α	[8]
Atorvastatin	Stroke model of rat intracerebral hemorrhage	Modulate neuroinflammation by dose-dependent reduction of TNF-α	[22]
Angiotensin-(1–7)	Stroke model of mice intracerebral hemorrhage	Modulate neuroinflammation by decrease in levels of TNF-α	[23]

Treatment	Experimental model	Related TNF finding	References
Milk fat globule-EGF factor VIII (MFG-E8)	Stroke model of rat permanent middle cerebral artery occlusion	Modulate neuroinflammation through decrease in expression of cerebral TNF-α level	[24]
Compound K (20-O-D-glucopyranosyl-20 (S)-protopanaxadiol)	Stroke model of mice transient middle cerebral artery occlusion	Modulate neuroinflammation through inhibition of lipopolysaccharide-induced production of TNF-α	[25]
Kaempferol glycosides	Stroke model of rat transient middle cerebral artery occlusion	Modulate neuroinflammation by inhibiting expression of tumor necrosis factor alpha	[26]
Angiotensin-(1–7)	Stroke model of rat permanent middle cerebral artery occlusion	Modulate neuroinflammation by inhibiting increase in TNF-α	[27]
Nicotine	Stroke model of rat global ischemia	Modulate neuroinflammation by reduction of enhanced expression of tumor necrosis factor alpha (TNF-α) induced by ischemia/reperfusion	[21]
Propofol	Brain neuroinflammation of lipopolysaccharide-induced inflammation in activated microglia	Modulate neuroinflammation by inhibiting lipopolysaccharide-mediated production TNF-α	[28]
Zileuton	Stroke model of rat permanent middle cerebral artery occlusion	Modulate neuroinflammation through attenuating release of TNF-α in the serum	[29]
Caffeic acid ester fraction (Caf)	Stroke model of rat middle cerebral artery occlusion in vivo and lipopolysaccharide-induced microglial activation in vitro	Modulate neuroinflammation by inhibiting TNF-α induced by lipopolysaccharide treatment in primary microglia in a dose-dependent manner	[30]
Telmisartan	Stroke model of rat intracerebral hemorrhage	Modulate neuroinflammation by decrease in tumor necrosis factor-α	[31]
Setarud (IMOD™)	Human patients with acute ischemic stroke	Modulate neuroinflammation by decrease in TNF-α levels	[32]
Caffeine	Brain neuroinflammation of lipopolysaccharide (LPS)-stimulated murine BV2 microglial cells	Modulate neuroinflammation by suppressing generation of proinflammatory TNF-α	[33]
SCH58261	Stroke model of rat bilateral common carotid artery occlusion	Modulate neuroinflammation by reversing ischemia reperfusion injury induced elevation of TNF-α.	[34]
Caffeine	Stroke model of rat bilateral common carotid artery occlusion	Modulate neuroinflammation by reduction of TNF-α activity	[35]
Fluoxetine	Stroke model of rat subarachnoid hemorrhage	Modulate neuroinflammation by decreasing the expression of proinflammatory mRNA levels of TNF-α	[36]
Matrix metalloproteinases 8 (MMP-8) inhibitor	Brain neuroinflammation of lipoteichoic acid (LTA)-stimulated rat primary astrocytes	Modulate neuroinflammation by inhibiting lipoteichoic acid (LTA) induced expression of TNF-α	[37]
Sildenafil	Brain neuroinflammation and demyelination induced by cuprizone in Mice model of multiple sclerosis.	Modulate neuroinflammation by reduction in the expression of the proinflammatory cytokines TNF-α	[38]

Table 1
Various agents that mediate neuroinflammation via TNF modulation

4. Agents that induce neuroinflammation

In a comprehensive review of agents that induce neuroinflammation, Nazeem [39] has classified models of neuroinflammation based on mechanism through which agents induce neuroinflammation into three as follows: immune challenge-based models which include lipopolysaccharide (LPS)-induced neuroinflammation and polyriboinosinic-polyribocytidilic acid (PolyI:C)-induced neuroinflammation; neurotoxin-induced models which consist of streptozotocin-induced neuroinflammation, okadaic acid-induced neuroinflammation, and colchicine-induced neuroinflammation; genetically manipulated models that contain interleukin-1β (IL-1β) overexpression model, p25 transgenic model, anti-nerve growth factor (NGF) transgenic models, and transforming growth factor-β (TGF-β)-deficient models.

The most commonly studied model of neuroinflammation is LPS-induced neuroinflammation which activates microglia in the brain [40]. LPS also termed endotoxin is a constituent of the external membrane of Gram-negative bacteria, and the mechanism of LPS-induced neuroinflammation is mediated through LPS binding with CD14 on microglia membranes. The LPS-CD14 complex then interacts with the Toll-like receptor-4 (TLR-4), which, in turn, activates microglia by initiating signal transduction cascades leading to rapid transcription and release of proinflammatory cytokines, chemokines, and the complement system proteins, as well as anti-inflammatory cytokines like IL-10 and transforming growth factor-β (TGF-β) [39].

Another popular emerging noninvasive, effective, and sterile method of induction neuroinflammation in animal model is MRI-guided pulsed focused ultrasound (pFUS) combined with systemic infusion of contrast agent microbubbles (MB). This MRI-guided pFUS+MB has advantage over all other methods of inducing neuroinflammation in a way that it induces neuroinflammation without systemic involvement [40].

5. Mechanism leading to the production of TNF-α in the brain and TNF-α signaling

Within the brain, TNF-α is produced and discharged in the brain predominantly by glial cells and neurons, with microglia and astrocytes being the major glial cells involved. Upon arrival of appropriate TNF-α production stimulus, TNF-α is formed as a 27-kDa (233 amino acids) precursor, which binds to cell membrane of producing cells. This precursor is cleaved by proteolysis to liberate a 17-kDa (157 amino acids) subunit by the action of TNF alpha-converting enzyme (TACE). TACE also known as ADAM17 is well-identified proteinase enzyme that mediates the process TNF-α production and is a member in the family of mammalian adamalysins (or ADAMs: A disintegrins and metalloproteinases) [41].

Upon cleavage by TACE/ADAM17, the free TNF-α forms a bioactive homotrimer that lead to biological effect of TNF-α. The actions of TNF-α is achieved through two distinct cell surface receptors: TNFR1 and TNFR2. TNF-α generates the activation of TNF receptors (TNFR1 and TNFR2), and the resultant TNF-induced TNFR signaling pathways are complex and wide ranging in different cell types, and precise circumstances, thereby accounting for TNF-α pleiotropic nature of action [5]. For instance, with TNFR1 signaling pathway, binding of TNF-α to the cognate receptor leads to the recruitment of TNF-α adaptor protein termed as TNF receptor-associated death domain (TRADD), which then creates a platform for binding of additional cytoplasmic adaptor proteins including TNF receptor-associated factor 2 (TRAF2), receptor-interacting protein (RIP), and FAS-associated death domain (FADD). The TRAF2 and RIP are concerned in escalating the transcriptional gene regulation; TRAF2 triggers the activation of a mitogen-activated protein kinase (MAPK) pathway, thereby leading to the activation of c-Jun N-terminal kinase (JNK), thus increasing its transcriptional activity; the RIP is a protein kinase vital to the activation of the transcription factor NFκB by phosphorylation of IκB kinase (IKK). On the other hand, FADD pathway leads to activation of caspase-8, thereby leading to initiate a caspase cascade of apoptosis cellular demise [41]. Although TNF-α binds to both TNFR1 and TNFR2 receptors with high affinity, there are some species specificity in terms of the receptor subtype and TNF-α binding [42]. TNF-α-induced p38 MAPK pathway transcription activity has been also implicated to induce proinflammatory IL-6 synthesis [43].

6. TNF-α and neuroinflammation

Neuroinflammation involves activation of microglia and astrocytes as well as influx of hematogenous cells recruited by cytokines, adhesion molecules, and chemokines across the activated blood vessel wall [44]. Neuroinflammatory signaling involves a coordinated effort of different molecules and cells types and is largely coordinated by a ubiquitous transcription factor NFκB. This signal transduction pathway for the activation of the transcription factor NFκB leads to control the expression of numerous genes activated during inflammation (i.e., cytokines, chemokines, growth factors, immune receptors, cellular ligands, and adhesion molecules). Thus, NFκB regulates a number of genes (including those coding for key inflammatory cytokines, like IL-6, TNF-α, etc.) involved in inflammation, making it the most important transcription factor that plays a key role in the inflammatory response. The collective gene targets of NFκB include various adhesion molecules, cytokines and chemokines (involved in proinflammatory signaling and NFκB activation, e.g., IL-1β and TNF-α), metalloproteinases (e.g., MMP-9), immune receptors, acute phase proteins, cell surface receptors, and inflammatory enzymes [45]. Various stimuli, such as cytokines, viruses, and oxidants, result in the activation of the transcription factor NFκB by separating it from inhibitor of NFκB alpha (IκBα)-bound protein in the cytoplasm, which becomes degraded and allows NFκB to move to the nucleus, where it binds to the DNA of the genes for numerous inflammatory mediators, resulting in their increased production and secretion [46].

It is pertinent to note that neuroinflammatory microglia-/macrophage-mediated phagocytosis is instrumental in neutralizing injurious foreign agent and conducting brain cleanup, the process which must occur to allow for tissue repair and functional recovery. This fast and efficient removal of apoptotic, dislocated, and damaged cells, before the discharge of injurious and proinflammatory cell contents occur, may help to reduce secondary damage. But inappropriate inflammatory responses generated by microglia/macrophages in a disease state may aggravate brain injury [45].

Proinflammatory TNF-α being one of the most key important early initiators of neuroinflammation interacts with two receptors R1 and R2, to mediate extrinsic apoptotic death signal via Fas-associated death domain (FADD) and inflammation via nuclear factor kappa-light-chain enhancer of activated B cells (NFκB), respectively [5]. NFκB is a major regulatory transcription factor with a pivotal role in inducing genes involved in inflammation [47]. In its dormant state, NFκB resides in the cytosol where it is bound to its inhibitory proteins known as inhibitors of NFκB (IκB), most commonly inhibitor of NFκB alpha (IκBα), making it unable to translocate into the nucleus [48]. Inflammatory stimuli resulting from wide range of brain pathological processes, such as cerebral ischemia, leads to degradation of these inhibitors upon their phosphorylation by the IκB kinase (IKK), which allows NFκB to migrate into the nucleus, where it binds with DNA, and activates transcription of many proinflammatory genes [49]. This includes increase in expression of the genes for proinflammatory cytokines, chemokines, enzymes that generate mediators of inflammation, and adhesion molecules [50]. Thus, TNF-α both activate and are activated by NFκB, creating a type positive regulatory loop that amplify and perpetuate local inflammation [50]. Hence, these pathways of TNF-α-induced NF-kB explain the ability of TNF-α to induce other inflammatory cytokines such as IL-6 and IL-8 and synergize with interferons [5].

Apart from IκB-NFκB pathway, another intracellular signaling pathway through which TNF-α induces other inflammatory cytokines is Janus family of tyrosine kinase (JAK)-signal transducer and activator of transcription (STAT) pathway. This JAK-STAT pathway can be initiated when there is TNF-α signaling after binding to its cognate receptors and consequently stimulates STATs. The STATs subsequently become activated and translocate to the nucleus to transmit transcriptional genetic expression of many cytokines, thereby leading to their synthesis [43].

Therefore, TNF-α is a proinflammatory cytokine that plays a critical role under both homeostatic and pathophysiological status within the central nervous system. Under healthy status, TNF-α has regulatory functions on vital physiological processes such as synaptic plasticity, learning and memory, sleep, food and water intake, and astrocyte-mediated synaptic amplification [51]. Under pathological status, astrocytes and mainly microglia excessively release massive concentration of TNF-α, thereby leading important constituent of neuroinflammatory response that marks a characteristic of several neurological disorders. Neuroinflammation itself at the first initial stage is a protective response in the brain, but excessively inappropriate inflammatory responses are detrimental, and in fact, it diminish the neuronal regeneration thereby leading to neurodegenerative diseases and other neurological disorders [52, 53].

7. Conclusion

Microglia is a pivotal brain endogenous protective mechanism against various injuries agents. If such an injury is tolerable, it triggers cellular responses that protect the brain and precondition the body against more severe stimuli. Beyond tolerable level, it triggers response that may potentially aggravate brain injury. TNF-α is released by microglia-induced NFκB activation, and activated NFκB in turn activates more TNF-α. The IκB-NFκB pathway together with other intracellular signaling pathway such as p38 MAPK pathway and JAK-STAT pathway that all orchestrate cascade of cytokine production makes TNF-α so-called master regulator of neuroinflammatory cytokine production. This phenomenon forms the basis of TNF-α as major cytokine of brain neuroinflammation.

Conflict of interest

The author declares no conflict of interest.

Author details

Mubarak Muhammad
Department of Human Physiology, Faculty of Basic Medical Sciences, College of Health Sciences, Bayero University Kano, Nigeria

*Address all correspondence to: mubarakmahmad@yahoo.com

References

[1] Carswell EA, Old LJ, Kassel RL, Green S, Fiore N, Williamson B. An endotoxin-induced serum factor that causes necrosis of tumors. Proceedings of the National Academy of Sciences of the United States of America. 1975; **72**(9):3666-3670

[2] Aggarwal BB, Gupta SC, Kim JH. Historical perspectives on tumor necrosis factor and its superfamily: 25 years later, a golden journey. Blood. 2012;**119**(3):651-665

[3] Sivangala R, Sumanlatha G. Cytokines that mediate and regulate immune responses. Innovative Immunology. 2015:1-26. https://www.austinpublishinggroup.com/ebooks/

[4] Commins SP, Borish L, Steinke JW. Immunologic messenger molecules: Cytokines, interferons, and chemokines. Journal of Allergy and Clinical Immunology. 2010;**125**:54-72

[5] Sedger LM, McDermottc MF. TNF and TNF-receptors: From mediators of cell death and inflammation to therapeutic giants—past, present and future. Cytokine and Growth Factor Reviews. 2014;**25**:453-472

[6] Denes A, Thornton P, Rothwell NJ, Allan SM. Inflammation and brain injury: Acute cerebral ischaemia, peripheral and central inflammation. Brain, Behavior, and Immunity. 2010; **24**:708-723

[7] Wang J, Yang Z, Liu C, Zhao Y, Chen Y. (2013). Activated microglia provide a neuroprotective role by balancing glial cell-line derived neurotrophic factor and tumor necrosis factor-α secretion after subacute cerebral ischemia. International Journal of Molecular Medicine. 2013;**31**(1):172-178

[8] Song JH, Lee JW, Shim B, Lee CY, Choi S, Kang C, et al. Glycyrrhizin alleviates neuroinflammation and memory deficit induced by systemic lipopolysaccharide treatment in mice. Molecules. 2013;**18**:15788-15803

[9] Xiong XY, Liang L, Yang QW. Functions and mechanisms of microglia/macrophages in neuroinflammation and neurogenesis during stroke. Progress in Neurobiology. 2016;**5**(1):1-108

[10] Walberer M, Jantzen SU, Backes H, Rueder MA, Keuters MH, Neumaier B, et al. In-vivo detection of inflammation and neurodegeneration in the chronic phase after permanent embolic stroke in rats. Brain Research. 2014;**1581**:180-188

[11] Lee Y, Lee SR, Choi SS, Yeo HG, Chang KT, Lee HJ. Therapeutically targeting neuroinflammation and microglia after acute ischemic stroke. BioMed Research International. 2014; **2014**:1-9

[12] Kalogeris T, Bao Y, Korthuis RJ. Mitochondrial reactive oxygen species: A double edged sword in ischemia/reperfusion vs preconditioning. Redox Biology. 2014;**2**:702-714

[13] Labro ME. Interference of antibacterial agents with phagocyte functions: Immunomodulation or "immuno-fairy tales"? Clinical Microbiology Reviews. 2000;**13**(4): 615-650

[14] Phaniendra A, Jestadi DB, Periyasamy L. Free radicals: Properties, sources, targets, and their implication in various diseases. Indian Journal of Clinical Biochemistry. 2015;**30**(1):11-26

[15] Onyango AN. Endogenous generation of singlet oxygen and ozone in human and animal tissues: Mechanisms, biological significance, and influence of dietary components. Oxidative Medicine and Cellular Longevity. 2016;**2016**:2398573

[16] Lakhan SE, Kirchgessner A, Hofer M. Inflammatory mechanisms in ischemic stroke: Therapeutic approaches. Journal of Translational Medicine. 2009;7(97):1-11

[17] Chang Y, Hsie CY, Peng, ZA, Yen TL, Hsiao G, Chou DS, 3 Chen CM, Sheu JR: Neuroprotective mechanisms of puerarin in middle cerebral artery occlusion-induced brain infarction in rats. Journal of Biomedical Science 2009; 16(9): 1–13.

[18] Yuan Y, Zha H, Ramgarajan P, Ling E, Wu C. Anti-inflammatory effects of Edaravone and Scutellarin in activated microglia in experimentally induced ischemia injury in rats and in BV-2 microglia. BMC Neuroscience. 2014;**15**: 125-133

[19] Han JU, Lee E, Moon E, Ryu JH, Choi JW, Kim H. Matrix metalloproteinase-8 is a novel pathogenetic factor in focal cerebral ischemia. Molecular Neurology. 2016; **53**(1):231-239

[20] Piao HZ, Jin SA, Chun HS, Lee J, Kim W. Neuro-protective effect of wogonin: Potential roles of inflammatory cytokines. Archives of Pharmacal Research. 2004;**27**(9): 930-936

[21] Guan Y, Jin X, Guan L, Yan H, Wang P, Gong Z, et al. Nicotine inhibits microglial proliferation and is neuro-protective in global ischemia rats. Molecular Neurobiology. 2015;**51**(3): 1480-1488

[22] Ewen T, Qiuting L, Chaogang T, Tao T, Jun W, Liming T, et al. Neuro-protective effect of atorvastatin involves suppression of TNF-α and upregulation of IL-10 in a rat model of intra-cerebral haemorrhage. Cell Biochemistry and Biophysics. 2013;**66**(2):337-346

[23] Bihl JC, Zhang C, Zhao Y, Xiao X, Ma X, Chen Y, et al. Angiotensin-(1-7) counteracts the effects of Ang II on vascular smooth muscle cells, vascular remodeling and hemorrhagic stroke: Role of the NFκB inflammatory pathway. Vascular Pharmacology. 2015; **73**:115-123

[24] Cheyuo C, Jacob A, Wu R, Zhou M, Qi L, Dong W, et al. Recombinant human MFG-E8 attenuates cerebral ischemic injury: Its role in anti-inflammation and anti-apoptosis. Neuropharmacology. 2012;**62**(2): 890-900

[25] Park JS, Shin JA, Jung JS, Hyun JW, Le TKV, Kim DH, et al. Anti-inflammatory mechanism of compound K in activated microglia and its neuroprotective effect on experimental stroke in mice. Journal of Pharmacology and Experimental Therapeutics. 2012; **341**(1):59-67

[26] Yu L, Chu Chen C, Wang LF, Kuang X, Liu K, Zhang H, et al. Neuroprotective effect of Kaempferol glycosides against brain injury and neuroinflammation by inhibiting the activation of NF-kB and STAT3 in transient focal stroke. PLoS One. 2013; **8**(2):1-11

[27] Jiang T, Gao L, Guo J, Lu J, Wang Y, Zhang Y. Suppressing inflammation by inhibiting the NF-kB pathway contributes to the neuroprotective effect of Angiotensin-(1-7) in rats with permanent cerebral ischaemia. British Journal of Pharmacology. 2012;**167**: 1520-1532

[28] Peng M, Ye JS, Wang Y, Chen C, Wang CY. Post treatment with Propofol attenuates lipopolysaccharide-induced up-regulation of inflammatory molecules in primary microglia. Inflammation Research. 2014;**63**(5): 411-418

[29] Tu K, Yang WZ, Wang CH, Shi SS, Chen CM, Yang YK, et al. Zileuton reduces inflammatory reaction and

brain damage following permanent cerebral ischemia in rats. Inflammation. 2010;**33**(5):344-352

[30] Wang SX, Guo H, Hu LM, Liu YN, Wang YF, Kang LY, et al. Caffeic acid ester fraction from *Erigeron breviscapus* inhibits microglial activation and provides neuroprotection. Chinese Journal of Integrative Medicine. 2012; **18**(6):437-444

[31] Jung KH, Chu K, Lee ST, Kim SJ, Song EC, Kim EH, et al. Blockade of AT1 receptor reduces apoptosis, inflammation, and oxidative stress in normotensive rats with intracerebral hemorrhage. Journal of Pharmacology and Experimental Therapeutics. 2007; **322**(3):1051-1058

[32] Farhoudi M, Najafi-Nesheli M, Hashemilar M, Mahmoodpoor A, Sharifipour E, Baradaran B, et al. Effect of IMOD™ on the inflammatory process after acute ischemic stroke: A randomized clinical trial. DARU Journal of Pharmaceutical Sciences. 2013; **21**(26):1-8

[33] Kanga C, Jayasooriyaa RG, Dilsharaa MG, Choib YH, Jeongc Y, Kimd ND, et al. Caffeine suppresses lipopolysaccharide-stimulated BV2 microglial cells by suppressing Akt-mediated NF-kB activation and ERK phosphorylation. Food and Chemical Toxicology. 2012;**50**:4270-4276

[34] Mohamed RA, Agha AM, Nassar NN. SCH58261 the selective adenosine A_{2A} receptor blocker modulates ischemia reperfusion injury following bilateral carotid occlusion: Role of inflammatory mediators. Neurochemical Research. 2012;**37**: 538-547

[35] Muhammad M, El-ta'alu AB, Mabrouk MA, Yarube IU, Nuhu JM, Yusuf I, et al. Effect of caffeine on serum tumour necrosis factor alpha and lactate dehydrogenase in Wistar rats exposed to cerebral ischaemia-reperfusion injury. Nigerian Journal of Physiological Sciences. 2018;**33**:001-008

[36] Liu FY, Cai J, Wang C, Ruan W, Guan GP, Pan HZ, et al. Fluoxetine attenuates neuroinflammation in early brain injury after subarachnoid hemorrhage: A possible role for the regulation of TLR4/MyD88/NF-κB signaling pathway. Journal of Neuroinflammation. 2018;**15**:347-459

[37] Lee EJ, Park JS, Lee YY, Kim DY, Kang JL, Kim HS. Anti-inflammatory and anti-oxidant mechanisms of an MMP-8 inhibitor in lipoteichoic acid-stimulated rat primary astrocytes: Involvement of NF-κB, Nrf2, and PPAR-γ signaling pathways. Journal of Neuroinflammation. 2018;**15**:326-338

[38] Nunes AKS, Raposo C, Rocha SWS, de Sousa Barbosa KP, de Almeidam RL, da Cruz-Höfling MA, et al. Involvement of AMP, IKβα-NFκB and eNOS in the sildenafil anti-inflammatory mechanism in a demyelination model. Brain Research. 2015;**1**(62):119-133

[39] Nazem A, Sankowski R, Bacher M, Al-Abed Y. Rodent models of neuroinflammation for Alzheimer's disease. Journal of Neuroinflammation. 2015;**12**:74

[40] Fung LK. A sterile animal model for neuroinflammation? Science Translational Medicine. 2017;**9**:373

[41] Figiel I. Pro-inflammatory cytokine TNF-α as a neuroprotective agent in the brain. Acta Neurobiologiae Experimentalis. 2008;**68**:526-534

[42] Parameswaran N, Patial S. Tumor necrosis factor-α signaling in macrophages. Critical Reviews in Eukaryotic Gene Expression. 2010; **20**(2):87-103

[43] Tanabe K, Matsushuma-Nishiwaki R, Yamaguchi S, Iida H, Dohi S, Kozawa O.

Mechanisms of tumor necrosis factor-α-induced interleukin-6 synthesis in glioma cells. Journal of Neuroinflammation. 2010;**7**:16

[44] Moskowitz MA, Lo EH, Iadecola C. The science of stroke: Mechanisms in search of treatments. Neuron. 2010; **67**(2):181-198

[45] Aronowski J, Zhao X. Molecular pathophysiology of cerebral hemorrhage: Secondary brain injury. Stroke. 2011;**42**(6):1781-1786

[46] Barrett K, Brooks H, Boitano S, Barman S. Cellular and molecular basis of medical physiology. In: Ganong's Review of Medical Physiology. 23rd ed. USA: McGraw-Hill Publishers; 2010. pp. 51-61

[47] Bowie A, O'Neill LA. Oxidative stress and nuclear factor-kB activation: A reassessment of the evidence in the light of recent discoveries. Biochemical Pharmacology. 2000;**59**:13-23

[48] Berlo DV, Knaapen AD, Schooten RP, Albrech C. (2010). NF-κB dependent and independent mechanisms of quartz-induced proinflammatory activation of lung epithelial cells. Particle and Fibre Toxicology. 2010;**7**(13):1-20

[49] Oeckinghaus A, Ghosh S. The NF-kB family of transcription factors and its regulation. Cold Spring Harbor Perspectives in Biology. 2009;**1**(4):1-14

[50] Barnes PJ, Karin M. Nuclear factor-kB: A pivotal transcription factor in chronic inflammatory diseases. The New England Journal of Medicine. 1997; **336**(15):1066-1071

[51] Olmos G, Llado J. Tumour necrosis factor alpha: A link between neuroinflammation and excitotoxicity. Mediators of Inflammation. 2014;**2014**: 861231

[52] Gresa-Arribas N, Vieitez C, Dentesano J, Saura J, Sola C. Modelling neuroinflammation in vitro: A tool to test the potential neuroprotective effect of anti-inflammatory agents. PLoS. 2012;**7**(9):e45227. DOI: 10.1371/journal.pone.0045227

[53] Kempuraj D, Thangavel R, Natteru PA, Selvakumar GP, Saeed D, Zahoor H, et al. Neuroinflammation induces neurodegeneration. Journal of Neurology Neurosurgery and Spine. 2016;**1**(1):1003

Chapter 4

TNFR2 and Regulatory T Cells: Potential Immune Checkpoint Target in Cancer Immunotherapy

Xuehui He and Xinhui Wang

Abstract

TNF has both proinflammatory and antiinflammatory effects. It binds to two structurally related but functionally distinct receptors TNFR1 and TNFR2. Unlike TNFR1 that is ubiquitously expressed, TNFR2 expression is more limited to myeloid and lymphoid cell lineages including a fraction of regulatory T cells (Treg). In general, TNFR1 is responsible for TNF-mediated cell apoptosis and death, and mostly induces proinflammatory reactions. However, TNFR2 mainly leads to functions related to cell survival and immune suppression. Treg play an indispensable role in maintaining immunological self-tolerance and restraining excessive immune reactions deleterious to the host. Impaired Treg-mediated immune regulation has been observed in various autoimmune diseases as well as in cancers. Therefore, Treg might provide an ideal therapeutic target for diseases where the immune balance is impaired and could benefit from the regulation of Treg properties. TNFR2 is highly expressed on Treg in mice and in humans, and TNFR2+ Treg reveal the most potent suppressive capacity. TNF-TNFR2 ligation benefits Treg proliferation, although the effect on Treg suppressive function remains controversial. Here, we will describe in detail the TNF-mediated regulation of Treg and the potential clinical applications in cancer immunotherapy as well as in autoimmune diseases, with the focus on human Treg subsets.

Keywords: TNF, TNF receptor 2, regulatory T cells, immunotherapy, autoimmune disease, cancer immunotherapy

1. Introduction

CD4+FOXP3+ regulatory T cells (Treg) have an indispensable role in maintaining immune homeostasis and immune tolerance. They control unwanted immune responses that are involved in the regulation of immune tolerance to self as well as to foreign antigens. Loss-of-function mutation in *FOXP3* locus, a gene encoding Treg lineage transcription factor FOXP3, leads to multiorgan associated autoimmunity. Abnormal numbers of Treg and/or impaired suppressive function of Treg are often found in various autoimmune diseases like type 1 diabetes (T1D) [1], multiple sclerosis (MS) [2], rheumatoid arthritis (RA) [3], psoriasis [4–6], and systemic lupus erythematosus (SLE) [7–9]. On the other hand, tumor-infiltrating Treg generally show potent suppressive functions, indicating that they regulate tumor-specific immune responses and might help tumor immune escape [10]. It seems logical to use Treg as a therapeutic target for diseases where the immune balance is impaired and could benefit from the regulation of Treg properties. Nevertheless, due to the intrinsic properties of Treg, i.e. heterogeneity and plasticity, several key questions need to be clarified before making Treg an ideal candidate for clinical applications.

Tumor necrosis factor (TNF) is initially expressed on cell surface as a membrane bound cytokine (mTNF), which can be cleaved by a metalloprotease TNF converting enzyme (TACE) to generate soluble form of TNF (sTNF) [11]. TNF binds to receptors, TNF receptor 1 (TNFR1) and 2 (TNFR2). In contrast to TNFR1, TNFR2 expression is restricted in certain cell types including lymphocytes [12]. TNF-TNFR1 interaction mostly induces proinflammatory reactions, whereas TNFR2 generally leads to the suppressive function of TNF [13]. It is known that TNFR2 is constitutively expressed on both murine and human Treg, and TNFR2+ Treg are the most suppressive Treg subpopulation [14–17]. The effect of TNF on Treg suppressor function remains controversial. In this chapter, we will describe in detail the TNF-mediated signal transduction pathways, its effect on Treg cells, and the potential clinical applications in various immunopathologies.

2. Regulatory T cells and its plasticity

Treg exert their function in primary and secondary lymphoid organs and non-lymphoid tissues. FOXP3, as the lineage transcription factor of Treg, facilitates Treg thymic development by stabilizing its own expression and inhibiting transcription factors needed for the development of other helper T-cell (Th) lineages like T-bet for Th1, GATA3 for Th2, and RORγt for Th17 cells [18]. Next to FOXP3, Treg constitutively express a high level of the IL-2 receptor α chain (CD25) and a low level of the IL-7 receptor α chain (CD127) compared to human activated non-Treg. The combination of CD4+, CD25high, and CD127low has been used to isolate Treg for functional studies and for adoptive immunotherapy [19]. However, no unique Treg marker has been identified so far, although many molecules are proposed. These Treg-related cell markers include CD27 [20], CD62L [21], CTLA4 (cytotoxic T-lymphocyte-associated protein) [22], CD39 and CD73 ectoenzymes [23], Helios [24], Neuropilin-1 [25], HLA-DR [26], and the most recently identified combination of TIGIT and FcRL3, which results in the identification of human Helios+ memory Treg [27].

Compelling evidence indicates that both mouse and human Treg consist of various subpopulations and have a more or less plastic phenotype depending on the microenvironment they are in [28]. Based on the site of Treg generation, two major Treg subsets are classified, namely, thymus-derived Treg (tTreg) that develop in the thymus from CD4 single positive thymocytes which in general display high-affinity self-reactive T-cell receptors (TCRs), and peripherally induced Treg (pTreg) which emerge in the periphery from conventional CD4+ T lymphocytes (Tconv) in response to environmental antigens and tolerogenic stimuli. Studies in mice have shown that pTreg and tTreg are both required for full protection against colitis and lymphoproliferative disease [29, 30], indicating that these two Treg subsets play distinct roles in protecting against immunopathology. However, the relative contribution of tTreg and pTreg in human immune tolerance remains a major unresolved issue, partially due to the lack of specific markers to definitively distinguish them. In fact, the transcription factor Helios was the first marker proposed to distinguish both mice and human tTreg from pTreg [31]. However, this has been disputed by studies showing that Helios can also be expressed by activated Tconv [32] and by pTreg upon in vitro and in vivo stimulation [33], precluding its use as tTreg-specific marker. Another cell surface marker that has been proposed to harbor the specificity necessary to distinguish between murine tTreg and pTreg is the coreceptor Neuropilin-1 [25]. Unfortunately, human Treg do not uniquely express Neuropilin-1 [34].

3. TNF/TNFR signaling pathways

TNF is firstly discovered as an inflammatory cytokine that is induced by the endotoxin [35]. Various immune cells produce TNF including macrophages, monocytes, dendritic cells, B cells, activated natural killer cells, and activated T cells. TNF is initially expressed on the cell surface as a trimeric type II transmembrane protein mTNF, which is then cleaved by the metalloproteinase TACE (also known as ADAM17) and released as soluble extracellular sTNF [36]. Both forms of TNF are present as bioactive homotrimers. There exist two structurally related but functionally distinct receptors, TNFR1 (p55) and TNFR2 (p75). TNFR1 is ubiquitously expressed on most mammalian cell types, and it binds to mTNF as well as sTNF, whereas TNFR2 expression is restricted to immune cells, neurons, and endothelial cells. TNFR2 binds with higher affinity to mTNF than sTNF compared to TNFR1.

TNFR1 and TNFR2 share the similar extracellular TNF-binding motifs but differ in their intracellular domains. Both receptors lack intrinsic enzyme activity; thus, upon the ligand binding, they need to recruit the cytosolic proteins to initiate the intracellular signal transduction. Specifically, TNFR1 contains a homologous intracellular region called "death domain", which preferentially interacts with the adaptor protein named TNFR1-associated death-domain (TRADD) protein [37]. TRADD further recruits another two adaptor proteins, receptor interacting protein kinase 1 (RIPK1) and TNFR-associated factor (TRAF) 2, thus forming an enzymatic complex signalosome, which is also known as signaling complex 1. One of the main targets of the complex 1 is the enzyme complex called IkB kinase (IKK). Phosphorylation of IKK in turn leads to the canonical activation of the transcription factor NFkB as well as members of the family of MAPKs such as c-jun kinase (JNK) and p38 MAPK. The TRADD containing signaling complex 1 may further be converted to a death-inducing signaling complex, so-called complex 2, by adaptor protein Fas-associated protein with death domain (FADD). The complex 2 is able to further initiate downstream caspase cascades, thus inducing cell apoptosis and cell death [37].

The pathways induced by TNFR2 are slightly different from TNFR1. Due to the lack of death domain, TNFR2 is unable to recruit TRADD protein, but it can directly interact with TRAF2 [38]. In contrast to TNFR1 that drives apoptosis and cell death, TNFR2 induces the noncanonical activation of NFκB via the activation of the NFκB-inducing kinase (NIK), which further leads to the phosphorylation of IKKα and the processing of p100, a crucial step in the nuclear translocation of p52/RelB [38, 39]. Interestingly, TRAF2 binding to TNFR2 is considerably weaker than its binding to TRADD protein. Upon binding to TRAF2, TNFR2 could also recruit cIAP1/2 proteins [39] that are involved in the TNFR1-mediated NFκB activation, indicating that there exists a crosstalk between TNFR1 and TNFR2 pathways. Another interesting adaptor protein called endothelial/epithelial protein tyrosine kinase (Etk) interacts with the C-terminal domain of TNFR2 in a ligand-independent manner [40]. TNFR2-mediated Etk phosphorylation is able to partially activate the growth factor receptor VEGFR2, which in turn results in the activation of PI3K/Akt pathway and cell survival.

A number of proteins are essential for the negative regulation of the TNF-TNFR pathways. A20, also named as TNF alpha-induced protein 3, is one of the most studied negative regulatory proteins. A20 is an ubiquitin editing enzyme. It limits NFκB signaling after activation by TNF [41]. Consistent with this, A20-deficient mice are hypersensitive to TNF exposure and die perinatally because of severe inflammation and multiorgan failure [42]. Intriguingly, A20 is recently shown to regulate the de novo generation of tTreg in a cell-intrinsic manner, while the suppressor function of A20-deficient Treg is unchanged in vitro [43].

4. Effect of TNFR2 on Treg

Although TNFR1 expression is not different between Treg and non-Treg cells, human Treg constitutively express high levels of TNFR2 compared to CD25- Tconv. Moreover, TNFR2+ Treg reveal the most potent suppressive capacity [14, 44]. The effect of TNF on Treg suppressor function remains controversial. Several groups including ours demonstrated that sTNF preserved or even increased FOXP3 expression as well as Treg suppressive capacity in both mice and humans [15, 45–47]. The TNF-TNFR2 is crucial for sustaining FOXP3 expression and maintaining the stability of murine Treg in an inflammatory environment [44]. A similar phenomenon is also observed for human Treg in vitro [48]. There is also evidence for the negative effects of TNF on Treg function. Studies show that TNF impairs Treg function by reducing FOXP3 expression or enhancing its dephosphorylation [47, 49]. In clinical practices, RA patients responding to anti-TNF antibody adalimumab showed an increased percentage of FOXP3 + cells as well as the restored regulatory function [50]. It should be noted that the nature of the TNFR2 antibodies used in these studies was likely different (agonistic versus antagonistic) [46]. Recent studies highlight that TNFR2 agonisms and antagonisms might regulate the phenotype and the suppressor function of Treg in a complete different way [46].

TNF priming induces the proliferation and activation of Treg in vitro [15, 51] as well as in vivo via TNFR2 in an acute mouse GvHD model [52]. Our group have found that stimulation of human Treg with a TNFR2-agonist antibody preserved a stable Treg phenotype and function after ex vivo expansion [48]. Using TNFR2 agonist only was enough to prevent the loss of FOXP3 expression, whereas the sustained hypomethylation of TSDR (Treg-specific demethylated region) of *FOXP3* gene locus required both rapamycin and TNFR2 agonist, suggesting that stabilization of FOXP3 expression requires both mTOR and NFκB signal pathways. In vitro restimulation of TNFR2 agonist plus rapamycin-expanded Treg led neither to the loss of FOXP3 protein nor the enhancement of IL-17A production, especially under proinflammatory conditions, indicating a well-preserved Treg stability. TNFR2 knockout CD4+ T cells have increased expression of RORγt and IL-17 production, which is dependent on the impairment of TNFR2-mediated activation of NFκB [53]. We speculate that a similar process of regulation may exist in human Treg where TNFR2/NFκB signaling might act as a double-edged sword to enhance FOXP3 but also to inhibit RORγt expression, thus contributing to Treg stability. Another possible explanation is that TNFR2 engagement results in an autocrine TNF-TNFR2 loop, which further regulates the expression of histone methyltransferase EZH2 [51], a subunit of the polycomb repressor complex 2 (PRC2). EZH2 is known to bind to FOXP3 thus helping FOXP3 to regulate the gene transcriptional repression [54].

5. TNFR2 agonists and autoimmune diseases

Defect in the function of Treg as well as the low numbers are the main properties of various autoimmune diseases. Therefore, restoring the proper functional Treg thus favoring the immune tolerance induction has become a final goal of treatment for patients with autoimmune diseases. As discussed above, ample studies show that either TNF and/or TNFR2 agonism has capacity to enhance Treg proliferation and activation. Furthermore, TNF-TNFR2 is essential to maintain the Treg function and stability in the inflammatory environment [44, 48]. Impaired TNF-TNFR signaling pathways occur in several human diseases including T1D, SLE, IBD, and MS. For instance, a single-nucleotide polymorphism (SNP) in the first intron is linked to a decreased level of TNFR2 in carriers of the SNP and a high risk of disease susceptibility [55]. T1D patients have higher TNFR2^{+} Treg compared to healthy controls. The rationale for using TNFR2 agonists as a therapeutic option for autoimmune diseases was first shown in T1D. Using blood from patients with T1D, a dose-response relationship between TNFR2 agonism and the destroying of pathogenic autoreactive CD8 T cells was observed [56], suggesting inducing of TNF-TNFR2 pathway is an effective approach of selectively killing autoreactive T cells.

Currently used biologics targeting TNF include the anti-TNF antibodies infliximab, adalimumab, certolizumab, and the decoy receptor etanercept that binds to sTNF. Although they have a good safety profile, with increasing use of these drugs, paradoxical adverse events involving the skin, joints, and lungs have been described [57]. Skin manifestations are the most common adverse event and occur in about 25% of patients receiving anti-TNFs. The underlying mechanism is recently attributed to the TNFR2/A20 signal axis which is specifically responsible for TNF-mediated IL-17A inhibition [58]. Termination of NFκB activation is critical to prevent aberrant inflammatory responses. In memory CD4 T cells, A20 is identified as one of the strongest TNF-responsive genes with a strong inverse correlation to IL-17A expression.

6. TNFR2 antagonists and cancer immunotherapy

Tumor microenvironment preferably recruits TNFR2+ Treg cells which possess a highly immunosuppressive capacity, thus facilitating tumor immune escape. That TNFR2 knockout mice show improved immune responses to tumors might be caused by the lack of TNFR2 expressing Treg or have failed to develop systemic autoimmunity [59] or the decreased numbers and the impaired function of MDSCs [60]. In humans, the high level of TNFR2+ Treg is found in the peripheral blood of lung cancer patients [10] and in the tumor-associated ascites in ovarian cancer patients [61]. Moreover, the increased TNFR2 gene expression on Treg cells has been shown to be associated with exhaustion of CD8 cytotoxic T lymphocytes in metastatic melanoma patients.

In addition to being an inducer of Treg expansion, TNFR2 also acts as an oncogene which has been identified on at least 25 tumor types. Enhanced expression of TNFR2 on tumor itself has been also reported but not limited in human renal cell carcinoma, multiple myeloma, colon cancer, ovarian cancer, and cutaneous T-cell lymphomas (CTCL) [62]. In general, the overexpression of TNFR2 exploits this cytokine receptor for increased tumor cell proliferation and tumor growth. Genetic mutation/genomic gains of *TNFRSF1B*, a gene encoding TNFR2 protein, occur in patients with Sézary syndrome (SS), a rare form of CTCL often refractory to treatment. SS is characterized with high expression of TNFR2 on the tumor cells and Treg. Such gain-of-function mutation in TNFR2 leads to the enhanced noncanonical NKκB activation [63], a pathway primarily involved in cell expansion and growth. It seems being desirable to apply one approach that could successfully inhibit potent suppressive Treg and also directly prevent tumor growth by using the antagonistic molecules against TNFR2. Such TNFR2-specific blocking molecules would ideally inhibit Treg and permit Tconv proliferation and function, thus enabling to restore the antitumor immune responses and to induce tumor regression.

7. Strategies for blocking of TNF/TNFR2 signaling

A number of agonistic or antagonistic biological agents targeting to TNF and/or TNFR2 have been developed. Two potent dominant TNFR2 antagonist antibodies are developed by Faustman et al. group [64]. They report that these TNFR2 antagonists lock the TNFR2 receptor in the form of antiparallel dimmers, which further prevents the TNF binding as well as the intracellular scaffolding. Consequently, these dominant TNFR2 antagonists, even in the presence of TNF, could kill Treg isolated from ovarian cancer ascites more potently than it kills Treg from healthy donors. Interestingly, TNFR2 antagonistic mAbs are also able to directly kill TNFR2-expression ovarian cancer cell lines in vitro [64]. Similar effect is observed in another in vitro study where the cancer cells and lymphocytes were isolated from the end-stage SS patients [65]. In mouse model of colon and breast cancers, combining a blocking TNFR2 antibody with a kind of immune stimulant markedly enhances the antitumor efficacy of immunotherapy through reducing the number of tumor-infiltrating TNFR2+ Treg and increasing the number of IFNγ-producing CD8 cells [66].

Some pharmacological agents are found to regulate TNF and/or its receptors expression. Thalidomide and its analogues prevent the surface expression of TNFR2 on activated T cells, which is associated with the inhibition of TNFR2 protein trafficking to the cell membrane [67]. Treating acute myeloid leukemia patients with azacitidine and lenalidomide, a thalidomide derivative can reduce TNFR2 expression on T cells as well as TNFR2+ Treg in vivo, leading to enhanced effector immune function [68]. Cyclophosphamide is a DNA alkylating agent. It is commonly used as a cytotoxic chemotherapy in cancer treatment. In a mouse model, it is shown that cyclophosphamide treatment depletes TNFR2+ Treg via inducing the death of replicating Treg that co-express TNFR2 and KI-67 [69]. A re-expansion of Treg from lymphodepletion suppresses the effective antitumor immunity developed after cyclophosphamide treatment. Intriguingly, blockade of TNF signaling using etanercept inhibits TNFR2+ Treg cell expansion during recovery from cyclophosphamide-induced lymphodepletion and markedly inhibits the growth of established CT26 tumors in mice [70]. Altogether, it suggests that a TNFR2-targeted approach to inactive host Treg, especially in only tumor microenvironment, may offer optimal options for antitumor immune reactions.

8. Conclusions

Many surface receptors of Treg are also expressed on other immune cells, with TNFR2 being a prominent exception with highest density in the tumor microenvironment. TNFR2 is a functional receptor on Treg. Cell surface expression of TNFR2 not only identifies the potent Treg subsets but also is the property of tumor-infiltrating Treg. TNFR2 expression on some cancer-infiltrating Treg is about 100 times higher than on circulating Treg in control subjects. In other types of cancer, the abundance of TNFR2+ Treg in peripheral blood is higher than healthy ones. Targeting TNFR2 using small molecule agonists or antagonists is a promising but also a challenging task. Considering the suppressive property of Treg and its impaired functions in various immunopathologies, there is no doubt that novel (tumor-specific) antagonists against TNFR2 are promising for cancer immunotherapy. From the clinical utilities point of view, combination of TNFR2 inhibition with immune checkpoint inhibitors seems to be an attractive approach in reshaping modern cancer immunotherapy.

Acknowledgements

The authors would like to thank the A FACTT network (Cost Action BM1305: http: //www.afactt.eu) for supporting this work by positive discussion. XH is also supported by NSFC 61263039 and NSFC 11101321. XW is supported by NSFC 61263039, NSFC 11101321, and 2018-ZJ-776.

The funders had no role in study design, data collection and analysis, decision to publish, or preparation of the manuscript.

Conflict of interest

The authors declare that the research was conducted in the absence of any commercial or financial relationships that could be construed as a potential conflict of interest.

Nomenclature

IBD	Inflammatory bowel disease
CTCL	Cutaneous T-cell lymphomas
MS	Multiple sclerosis
MAPK	Mitogen-activated protein kinase
mTNF	Membrane-bound TNF
NFκB	Nuclear factor κB
RA	Rheumatoid arthritis
SNP	Single-nucleotide polymorphism
SS	Sézary syndrome
T1D	Type 1 diabetes
TACE	TNF-converting enzyme
TCR	T-cell receptor
TNFR	TNF receptor
TRAF	TNFR-associated factor
Treg	Regulatory T cells
TSDR	Treg-specific demethylated region

Author details

Xuehui He[1,2]* and Xinhui Wang[1,3]

1 College of Computer Science, Qinghai Normal University, Xining, Qinghai, China

2 Department of Laboratory Medicine, Laboratory Medical Immunology, Radboud University Medical Center, Nijmegen, The Netherlands

3 Department of Public and Occupational Health, Amsterdam Public Health Research Institute, Amsterdam University Medical Center, Amsterdam, The Netherlands

*Address all correspondence to: xuehui.he@radboudumc.nl

References

[1] Bluestone JA, Buckner JH, Fitch M, Gitelman SE, Gupta S, Hellerstein MK, et al. Type 1 diabetes immunotherapy using polyclonal regulatory T cells. Science Translational Medicine. 2015;7(315):315ra189

[2] Haas J, Fritzsching B, Trubswetter P, Korporal M, Milkova L, Fritz B, et al. Prevalence of newly generated naive regulatory T cells (Treg) is critical for Treg suppressive function and determines Treg dysfunction in multiple sclerosis. Journal of Immunology. 2007;**179**(2):1322-1330

[3] Komatsu N, Okamoto K, Sawa S, Nakashima T, Oh-hora M, Kodama T, et al. Pathogenic conversion of Foxp3+ T cells into TH17 cells in autoimmune arthritis. Nature Medicine. 2014;**20**(1):62-68

[4] Bovenschen HJ, van de Kerkhof PC, van Erp PE, Woestenenk R, Joosten I, Koenen HJ. Foxp3+ regulatory T cells of psoriasis patients easily differentiate into IL-17A-producing cells and are found in lesional skin. Joural of Investigative Dermatology. 2011;**131**(9):1853-1860

[5] Sugiyama H, Gyulai R, Toichi E, Garaczi E, Shimada S, Stevens SR, et al. Dysfunctional blood and target tissue CD4+CD25 high regulatory T cells in psoriasis: Mechanism underlying unrestrained pathogenic effector T cell proliferation. Journal of Immunology. 2005;**174**(1):164-173

[6] Keijsers RR, van der Velden HM, van Erp PE, de Boer-van Huizen RT, Joosten I, Koenen HJ, et al. Balance of Treg vs. T-helper cells in the transition from symptomless to lesional psoriatic skin. The British Journal of Dermatology. 2013;**168**(6):1294-1302

[7] Ohl K, Tenbrock K. Regulatory T cells in systemic lupus erythematosus. European Journal of Immunology. 2015;**45**(2):344-355

[8] Lyssuk EY, Torgashina AV, Soloviev SK, Nassonov EL, Bykovskaia SN. Reduced number and function of CD4+CD25 high FoxP3+ regulatory T cells in patients with systemic lupus erythematosus. Advances in Experimental Medicine and Biology. 2007;**601**:113-119

[9] Wildin RS, Ramsdell F, Peake J, Faravelli F, Casanova JL, Buist N, et al. X-linked neonatal diabetes mellitus, enteropathy and endocrinopathy syndrome is the human equivalent of mouse scurfy. Nature Genetics. 2001;**27**(1):18-20

[10] Yan F, Du R, Wei F, Zhao H, Yu J, Wang C, et al. Expression of TNFR2 by regulatory T cells in peripheral blood is correlated with clinical pathology of lung cancer patients. Cancer Immunology, Immunotherapy: CII. 2015;**64**(11):1475-1485

[11] Moss ML, Jin SL, Milla ME, Bickett DM, Burkhart W, Carter HL, et al. Cloning of a disintegrin metalloproteinase that processes precursor tumour-necrosis factor-alpha. Nature. 1997;**385**(6618):733-736

[12] Grell M, Douni E, Wajant H, Lohden M, Clauss M, Maxeiner B, et al. The transmembrane form of tumor necrosis factor is the prime activating ligand of the 80 kDa tumor necrosis factor receptor. Cell. 1995;**83**(5):793-802

[13] Kim EY, Priatel JJ, Teh SJ, Teh HSTNF. Receptor type 2 (p75) functions as a costimulator for antigen-driven T cell responses in vivo. Journal of Immunology (Baltimore Md: 1950). 2006;**176**(2):1026-1035

[14] Chen X, Subleski JJ, Hamano R, Howard OM, Wiltrout RH, Oppenheim

JJ. Co-expression of TNFR2 and CD25 identifies more of the functional CD4+FOXP3+ regulatory T cells in human peripheral blood. European Journal of Immunology. 2010;**40**(4):1099-1106

[15] Chen X, Baumel M, Mannel DN, Howard OM, Oppenheim JJ. Interaction of TNF with TNF receptor type 2 promotes expansion and function of mouse CD4+CD25+ T regulatory cells. Journal of Immunology (Baltimore Md: 1950). 2007;**179**(1):154-161

[16] Atarashi K, Tanoue T, Shima T, Imaoka A, Kuwahara T, Momose Y, et al. Induction of colonic regulatory T cells by indigenous clostridium species. Science. 2011;**331**(6015):337-341

[17] Annunziato F, Cosmi L, Liotta F, Lazzeri E, Manetti R, Vanini V, et al. Phenotype, localization, and mechanism of suppression of CD4(+) CD25(+) human thymocytes. The Journal of Experimental Medicine. 2002;**196**(3):379-387

[18] Josefowicz SZ, Lu LF, Rudensky AY. Regulatory T cells: Mechanisms of differentiation and function. Annual Review of Immunology. 2012;**30**:531-564

[19] Hartigan-O'Connor DJ, Poon C, Sinclair E, McCune JM. Human CD4+ regulatory T cells express lower levels of the IL-7 receptor alpha chain (CD127), allowing consistent identification and sorting of live cells. Journal of Immunological Methods. 2007;**319**(1-2):41-52

[20] Koenen HJ, Joosten I. Antigen-specific regulatory T-cell subsets in transplantation tolerance regulatory T-cell subset quality reduces the need for quantity. Human Immunology. 2006;**67**(9):665-675

[21] Ermann J, Hoffmann P, Edinger M, Dutt S, Blankenberg FG, Higgins JP, et al. Only the CD62L+ subpopulation of CD4+CD25+ regulatory T cells protects from lethal acute GVHD. Blood. 2005;**105**(5):2220-2226

[22] Wing K, Onishi Y, Prieto-Martin P, Yamaguchi T, Miyara M, Fehervari Z, et al. CTLA-4 control over Foxp3+ regulatory T cell function. Science. 2008;**322**(5899):271-275

[23] Ohta A, Sitkovsky M. Extracellular adenosine-mediated modulation of regulatory T cells. Frontiers in Immunology. 2014;**5**:304

[24] Himmel ME, MacDonald KG, Garcia RV, Steiner TS, Levings MK. Helios+ and Helios- cells coexist within the natural FOXP3+ T regulatory cell subset in humans. Journal of Immunology. 2013;**190**(5):2001-2008

[25] Bruder D, Probst-Kepper M, Westendorf AM, Geffers R, Beissert S, Loser K, et al. Neuropilin-1: A surface marker of regulatory T cells. European Journal of Immunology. 2004;**34**(3):623-630

[26] Baecher-Allan C, Wolf E, Hafler DA. MHC class II expression identifies functionally distinct human regulatory T cells. Journal of Immunology. 2006;**176**(8):4622-4631

[27] Bin Dhuban K, d'Hennezel E, Nashi E, Bar-Or A, Rieder S, Shevach EM, et al. Coexpression of TIGIT and FCRL3 identifies Helios+ human memory regulatory T cells. Journal of Immunology. 2015;**194**(8):3687-3696

[28] Li X, Zheng Y. Regulatory T cell identity: Formation and maintenance. Trends in Immunology. 2015;**36**(6):344-353

[29] Josefowicz SZ, Niec RE, Kim HY, Treuting P, Chinen T, Zheng Y, et al. Extrathymically generated regulatory T cells control mucosal TH2 inflammation. Nature. 2012;**482**(7385):395-399

[30] Cebula A, Seweryn M, Rempala GA, Pabla SS, McIndoe RA, Denning TL, et al. Thymus-derived regulatory T cells contribute to tolerance to commensal microbiota. Nature. 2013;**497**(7448):258-262

[31] Thornton AM, Korty PE, Tran DQ, Wohlfert EA, Murray PE, Belkaid Y, et al. Expression of Helios, an Ikaros transcription factor family member, differentiates thymic-derived from peripherally induced Foxp3+ T regulatory cells. Journal of Immunology. 2010;**184**(7):3433-3441

[32] Akimova T, Beier UH, Wang L, Levine MH, Hancock WW. Helios expression is a marker of T cell activation and proliferation. PLoS One. 2011;**6**(8):e24226

[33] Gottschalk RA, Corse E, Allison JP. Expression of Helios in peripherally induced Foxp3+ regulatory T cells. Journal of Immunology. 2012;**188**(3):976-980

[34] Milpied P, Renand A, Bruneau J, Mendes-da-Cruz DA, Jacquelin S, Asnafi V, et al. Neuropilin-1 is not a marker of human Foxp3+ Treg. European Journal of Immunology. 2009;**39**(6):1466-1471

[35] Carswell EA, Old LJ, Kassel RL, Green S, Fiore N, Williamson B. An endotoxin-induced serum factor that causes necrosis of tumors. Proceedings of the National Academy of Sciences of the United States of America. 1975;**72**(9):3666-3670

[36] Pennica D, Nedwin GE, Hayflick JS, Seeburg PH, Derynck R, Palladino MA, et al. Human tumour necrosis factor: Precursor structure, expression and homology to lymphotoxin. Nature. 1984;**312**(5996):724-729

[37] Dempsey PW, Doyle SE, He JQ, Cheng G. The signaling adaptors and pathways activated by TNF superfamily. Cytokine & Growth Factor Reviews. 2003;**14**(3-4):193-209

[38] Rothe M, Pan MG, Henzel WJ, Ayres TM, Goeddel DV. The TNFR2-TRAF signaling complex contains two novel proteins related to baculoviral inhibitor of apoptosis proteins. Cell. 1995;**83**(7):1243-1252

[39] Borghi A, Haegman M, Fischer R, Carpentier I, Bertrand MJM, Libert C, et al. The E3 ubiquitin ligases HOIP and cIAP1 are recruited to the TNFR2 signaling complex and mediate TNFR2-induced canonical NF-kappaB signaling. Biochemical Pharmacology. 2018;**153**:292-298

[40] Pan S, An P, Zhang R, He X, Yin G, Min W. Etk/Bmx as a tumor necrosis factor receptor type 2-specific kinase: Role in endothelial cell migration and angiogenesis. Molecular and Cellular Biology. 2002;**22**(21):7512-7523

[41] Catrysse L, Vereecke L, Beyaert R, van Loo G. A20 in inflammation and autoimmunity. Trends in Immunology. 2014;**35**(1):22-31

[42] Lee EG, Boone DL, Chai S, Libby SL, Chien M, Lodolce JP, et al. Failure to regulate TNF-induced NF-kappaB and cell death responses in A20-deficient mice. Science (New York NY). 2000;**289**(5488):2350-2354

[43] Fischer JC, Otten V, Kober M, Drees C, Rosenbaum M, Schmickl M, et al. A20 restrains thymic regulatory T cell development. Journal of Immunology (Baltimore Md: 1950). 2017;**199**(7):2356-2365

[44] Chen X, Wu X, Zhou Q, Howard OM, Netea MG, Oppenheim JJ. TNFR2 is critical for the stabilization of the CD4+Foxp3+ regulatory T. Cell phenotype in the inflammatory environment. Journal of Immunology (Baltimore Md: 1950). 2013;**190**(3):1076-1084

[45] Zaragoza B, Chen X, Oppenheim JJ, Baeyens A, Gregoire S, Chader D, et al. Suppressive activity of human regulatory T cells is maintained in the presence of TNF. Nature Medicine. 2016;**22**(1):16-17

[46] Okubo Y, Mera T, Wang L, Faustman DL. Homogeneous expansion of human T-regulatory cells via tumor necrosis factor receptor 2. Scientific Reports. 2013;**3**:3153

[47] Nie H, Zheng Y, Li R, Guo TB, He D, Fang L, et al. Phosphorylation of FOXP3 controls regulatory T cell function and is inhibited by TNF-alpha in rheumatoid arthritis. Nature Medicine. 2013;**19**(3):322-328

[48] He X, Landman S, Bauland SC, van den Dolder J, Koenen HJ, Joosten I. A TNFR2-agonist facilitates high purity expansion of human low purity Treg cells. PLoS One. 2016;**11**(5):e0156311

[49] Valencia X, Stephens G, Goldbach-Mansky R, Wilson M, Shevach EM, Lipsky PE. TNF downmodulates the function of human CD4+CD25hi T-regulatory cells. Blood. 2006;**108**(1):253-261

[50] McGovern JL, Nguyen DX, Notley CA, Mauri C, Isenberg DA, Ehrenstein MR. Th17 cells are restrained by Treg cells via the inhibition of interleukin-6 in patients with rheumatoid arthritis responding to anti-tumor necrosis factor antibody therapy. Arthritis and Rheumatism. 2012;**64**(10):3129-3138

[51] Urbano PCM, Koenen H, Joosten I, He X. An autocrine TNFalpha-tumor necrosis factor receptor 2 loop promotes epigenetic effects inducing human Treg stability in vitro. Frontiers in Immunology. 2018;**9**:573

[52] Leclerc M, Naserian S, Pilon C, Thiolat A, Martin GH, Pouchy C, et al. Control of GVHD by regulatory T cells depends on TNF produced by T cells and TNFR2 expressed by regulatory T cells. Blood. 2016;**128**(12):1651-1659

[53] Miller PG, Bonn MB, McKarns SC. Transmembrane TNF-TNFR2 impairs Th17 differentiation by promoting Il2 expression. Journal of Immunology (Baltimore Md: 1950). 2015;**195**(6):2633-2647

[54] Arvey A, van der Veeken J, Samstein RM, Feng Y, Stamatoyannopoulos JA, Rudensky AY. Inflammation-induced repression of chromatin bound by the transcription factor Foxp3 in regulatory T cells. Nature Immunology. 2014;**15**(6):580-587

[55] Li D, Silverberg MS, Haritunians T, Dubinsky MC, Landers C, Stempak JM, et al. TNFRSF1B is associated with ANCA in IBD. Inflammatory Bowel Diseases. 2016;**22**(6):1346-1352

[56] Ban L, Zhang J, Wang L, Kuhtreiber W, Burger D, Faustman DL. Selective death of autoreactive T cells in human diabetes by TNF or TNF receptor 2 agonism. Proceedings of the National Academy of Sciences of the United States of America. 2008;**105**(36):13644-13649

[57] Cleynen I, Vermeire S. Paradoxical inflammation induced by anti-TNF agents in patients with IBD. Nature Reviews Gastroenterology and Hepatology. 2012;**9**(9):496-503

[58] Urbano PCM, Aguirre-Gamboa R, Ashikov A, van Heeswijk B, Krippner-Heidenreich A, Tijssen H, et al. TNF-alpha-induced protein 3 (TNFAIP3)/A20 acts as a master switch in TNF-alpha blockade-driven IL-17A expression. The Journal of Allergy and Clinical Immunology. 2018;**142**(2):517-529

[59] Chen X, Hamano R, Subleski JJ, Hurwitz AA, Howard OM, Oppenheim JJ. Expression of costimulatory TNFR2

induces resistance of CD4+FoxP3- conventional T cells to suppression by CD4+FoxP3+ regulatory T cells. Journal of Immunology (Baltimore Md: 1950). 2010;**185**(1):174-182

[60] Polz J, Remke A, Weber S, Schmidt D, Weber-Steffens D, Pietryga-Krieger A, et al. Myeloid suppressor cells require membrane TNFR2 expression for suppressive activity. Immunity, Inflammation and Disease. 2014;**2**(2):121-130

[61] Govindaraj C, Scalzo-Inguanti K, Madondo M, Hallo J, Flanagan K, Quinn M, et al. Impaired Th1 immunity in ovarian cancer patients is mediated by TNFR2+ Tregs within the tumor microenvironment. Clinical Immunology. 2013;**149**(1):97-110

[62] Vanamee ES, Faustman DL. TNFR2: A novel target for cancer immunotherapy. Trends in Molecular Medicine. 2017;**23**(11):1037-1046

[63] Ungewickell A, Bhaduri A, Rios E, Reuter J, Lee CS, Mah A, et al. Genomic analysis of mycosis fungoides and Sezary syndrome identifies recurrent alterations in TNFR2. Nature Genetics. 2015;**47**(9):1056-1060

[64] Torrey H, Butterworth J, Mera T, Okubo Y, Wang L, Baum D, et al. Targeting TNFR2 with antagonistic antibodies inhibits proliferation of ovarian cancer cells and tumor-associated Tregs. Science Signaling. 2017;**10**(462). pii: eaaf8608

[65] Torrey H, Khodadoust M, Tran L, Baum D, Defusco A, Kim YH, et al. Targeted killing of TNFR2-expressing tumor cells and Tregs by TNFR2 antagonistic antibodies in advanced Sezary syndrome. Leukemia. 2018 Oct 24. pii: 10.1038/s41375-018-0292-9

[66] Nie Y, He J, Shirota H, Trivett AL, Yang KDM, et al. Blockade of TNFR2 signaling enhances the immunotherapeutic effect of CpG ODN in a mouse model of colon cancer. Science Signaling. 2018;**11**(511). pii: 11/511/eaan0790

[67] Marriott JB, Clarke IA, Dredge K, Muller G, Stirling D, Dalgleish AG. Thalidomide and its analogues have distinct and opposing effects on TNF-alpha and TNFR2 during co-stimulation of both CD4(+) and CD8(+) T cells. Clinical and Experimental Immunology. 2002;**130**(1):75-84

[68] Govindaraj C, Madondo M, Kong YY, Tan P, Wei A, Plebanski M. Lenalidomide-based maintenance therapy reduces TNF receptor 2 on CD4 T cells and enhances immune effector function in acute myeloid leukemia patients. American Journal of Hematology. 2014;**89**(8):795-802

[69] van der Most RG, Currie AJ, Mahendran S, Prosser A, Darabi A, Robinson BW, et al. Tumor eradication after cyclophosphamide depends on concurrent depletion of regulatory T cells: A role for cycling TNFR2-expressing effector-suppressor T cells in limiting effective chemotherapy. Cancer Immunology, Immunotherapy: CII. 2009;**58**(8):1219-1228

[70] Chang LY, Lin YC, Chiang JM, Mahalingam J, Su SH, Huang CT, et al. Blockade of TNF-alpha signaling benefits cancer therapy by suppressing effector regulatory T cell expansion. Oncoimmunology. 2015;**4**(10):e1040215

Chapter 5

Innate Immunity and Neuroinflammation in Neuropsychiatric Conditions Including Autism Spectrum Disorders: Role of Innate Immune Memory

Harumi Jyonouchi

Abstract

The neuroimmune network represents a dense network of multiple signals mediated by neurotransmitters, hormones, growth factors, and cytokines produced by multiple lineage cells and is crucial for maintaining neuroimmune homeostasis. Endogenous and exogenous stimuli, which are dangerous to the body, are detected by sensor cells, and they rapidly inform the brain through this network. Innate immunity is thought to play a major role in the neuroimmune network, through cytokines and other mediators released from secretary innate immune cells. Recent research has revealed that innate immunity has its own memory. This is accomplished by metabolic and epigenetic changes. Such changes may result in augmenting immune protection with a risk of excessive inflammatory responses to subsequent stimuli (trained immunity). Alternatively, innate immune memory can induce suppressive effects (tolerance), which may impose a risk of impaired immune defense. Innate immune memory affects the neuroimmune network for a prolonged period, and dysregulated innate immune memory has been implicated with pathogenesis of neuropsychiatric conditions. This chapter summarizes a role of innate immune memory (trained immunity vs. tolerance) in neuroinflammation in association with neuropsychiatric conditions including autism spectrum disorders (ASD).

Keywords: innate immunity, cytokines, neuroinflammation, neuroimmune network, immune metabolic processes

1. Introduction

It is well accepted that inflammation in the peripheral organs can influence homeostasis and immune responses in the central nervous system (CNS) [1]. In common neuropsychiatric conditions such as schizophrenia and depression, evidence indicates that neuroinflammation plays a role in the disease pathogenesis [2]. Long-lasting effects of neuroinflammation in such neuropsychiatric conditions are implicated with altered innate immune responses in the absence of specific pathogens [2]. However, until recently, it is not well understood how innate immunity, which was thought to have no lasting memory unlike adaptive immunity, can exert prolonged actions on the CNS. The recent discovery of innate immune memory (trained immunity vs. tolerance) shed a light in a long postulated role of innate immunity in neuropsychiatric diseases [3, 4].

Since the existence of the immune system was recognized more than 50 years ago, the immune system has been thought to be comprised of two components, innate immunity and adaptive immunity. Innate immunity is the arm that mounts nonspecific, acute immune responses, by sensing microbial by-products called pathogen-associated molecular patterns (PAMPs) or by-products derived from tissue injuries called damage-associated molecular patterns (DAMPs) [5]. Signaling through PAMPs and DAMPs are thought to play a major role in plant immunity [6]. In animals, adaptive immunity is the arm that develops antigen (Ag)-specific responses. The development of Ag-specific responses requires lengthy processes including antigen (Ag) processing by Ag-presenting cells (APCs), Ag presentation to T and B cells, and TCR or immunoglobulin gene arrangements of T and B lymphocytes, respectively, which lead to the development of Ag-specific T and B cells and finally antibodies (Abs) [7]. Adaptive immunity effectively eliminates hazards from the body through Ag-specific cellular and humoral immune responses [7]. Adaptive immunity results in the development of long-lasting Ag-specific memory T/B cells [8]. In this way, the body retains immune memory against specific pathogens for a prolonged time. It is well known that individuals who have survived measles will retain measles-specific immune defense for life.

In contrast, immunology textbooks have long taught us that innate immunity does not have any lasting effects or memory, and it is mainly effective in containing infection until adaptive immunity takes over. Innate immunity has also been known to shape adaptive immunity through multiple mechanisms such as affecting actions of APCs, thereby indirectly modifying adaptive immune responses [7]. However, recent exciting research revealed that innate immunity can have its own memory, following an immune stimulus, and this depends on time, amount, and the kinds of stimuli through metabolic and epigenetic changes [3, 9]. More importantly, the stimuli that evoke innate immune memory are not restricted to microbes; nonpathogenic challenges such as stress and obesity are also found to cause innate immune memory [3, 10].

As described previously, despite the accumulating evidence, it was difficult to understand how innate immunity exerts lasting effects, in the absence of specific pathogens or other persistent environmental stimuli, in neuropsychiatric conditions. The recognition of innate immune memory (trained immunity vs. tolerance) has provided us new insights with regard to the role of innate immunity in physiological as well as pathogenic consequences in the brain. In this chapter, research efforts shaping a concept of innate immune memory (trained immunity vs. innate immune tolerance) will be discussed first. In the latter part of the chapter, a potential role of innate immune memory in neuropsychiatric conditions, especially in ASD, will be discussed.

2. Innate immune memory

2.1 Trained immunity

The presence of innate immune memory was first suspected because of unexpected, nonspecific effects of vaccinations. This is best known for a Bacillus

Calmette-Guérin (BCG) vaccine. Epidemiological studies and subsequent randomized trials showed that the BCG vaccination not only provided protection for tuberculosis but also protection against other pathogens, especially those causing respiratory infection, which resulted in a reduction in infant mortality greater than expected for reducing tuberculosis-associated mortality [11, 12]. Likewise, the measles vaccination resulted in a striking reduction in children's mortality, which was again not to be explained by the reduction in mortality caused by measles [11]. These epidemiological observations were further explored by researchers in the Netherlands. They first demonstrated that innate immune memory does exist in animal models [13]. Namely, these researchers showed that BCG provided enhanced protection against *Candida albicans* through nonspecific adaptation of innate immunity, independent of lymphocytes [13]. They proposed to name this process of innate immune memory "trained immunity." The following studies by the same group also revealed that such adaptive changes in innate immunity are present not only in monocyte-macrophage lineage cells but also in other innate immune cells such as natural killer (NK) cells [14] and progenitor cells of innate immune cells in the bone marrow [15, 16]. Further studies revealed the presence of trained immunity in humans [17–19]. It became clear that trained immunity is similar to plant immunity which does not develop Ag-specific immunity, but develops prolonged immune defense by metabolic and epigenetic modulation [20]. Mounting evidence has now repeatedly shown that trained immunity is Ag nonspecific; the second stimulus (DAMP or PAMP) causing innate immune activation can be different from the first stimulus [3].

2.2 Mechanisms of trained immunity

Adaptive changes observed in "in vitro" models of trained immunity with β-glucan, a representative PAMP from *Candida albicans*, have been extensively studied. It was revealed that ß-glucan treatment induces activation of the dectin-1/Akt/PTEN/mTOR/HIF-1α signaling pathway in innate immune cells [21]. That is, β-glucan activates dectin-1 which recruits Akt, leading to activation of mammalian target of rapamycin (mTOR) with suppression of PTEN expression and phosphorylation of the tuberous sclerosis complex (TSC) [22]. Activation of this pathway switches cellular metabolism from oxidative phosphorylation (ATP synthesis) to glycolysis, thereby reducing basal cellular respiration and increasing in glucose consumption, resulting in higher production of lactate [21]. Such metabolic changes lead to the exportation of citrate to the cytoplasm for cholesterol synthesis and phospholipid synthesis [23, 24].

This metabolic shift described above results in the replenishment of the Krebs cycle by metabolization of glutamine into glutamate and α-keto-glutamate, leading to an accumulation of fumarate [23, 24]. Higher concentration of fumarate inhibits the KDM5 family of H3K4 demethylase that eventually leads to epigenetic reprogramming [23]. It has been reported that in the initial phase of trained immunity, lysine 27 of histone 3 (H3K27) is acetylated and lysine 4 of histone 3 (H3K4) is methylated rapidly [25]. Although H3K27Ac gradually returns to the baseline over time, H3K4me3 was found to remain elevated in the trained immunity [25]. Such epigenetic histone modification (accumulation of H3K4me3) is known to lead to the remodeling of the local chromatin into an open and accessible state, resulting in the facilitation of the loading of transcriptional machinery. The remaining accumulation of H3K4me3 on chromatin has been implicated in the establishment of the epigenetic memory in the trained immunity [25, 26]. It was hypothesized that H3K4me3 increases the local hydrophobicity of the chromatin, allowing for liquid-liquid phase separated transcription factors to engage with the DNA in the aqueous environment of the nucleus, subsequently rendering loading of transcriptional machinery onto promoters [27–29]. This will allow cells to start rapid transcription of the genes necessary for immune responses, thereby causing a much stronger Ag nonspecific pro-inflammatory response.

Long noncoding RNAs (lncRNAs) can function as a molecular scaffold where multiple protein complexes can assemble, and they also guide these complexes to specific gene loci [30]. Recent research disclosed a new class of lncRNAs named immune gene-priming lncRNAs (IPLs), and IPLs were found to have a crucial role in the accumulation of H3K4me3 on chromatin [31]. A candidate IPL, termed upstream master lncRNA of the inflammatory chemokine locus (UMLILO), was found to be crucial for trained immunity; ablation of the UMLILO transcript abolished β-glucan-induced trained immunity in both human and murine monocytes [30].

As shown in epidemiological studies of vaccinations, trained immunity, caused by metabolic and epigenetic changes, will be beneficial in providing broader immune defense and promoting tissue repair [32]. On the other hand, maladapted trained immunity can be detrimental to human health. Chronic inflammatory conditions including neuropsychiatric conditions have been implicated with maladapted changes in trained immunity [2, 9]. It should also be noted that induction of trained immunity appears to be associated with doses of PAMP, perhaps DAMP in humans; depending on the dose and the kinds of PAMP/DAMP, tolerance can be induced, instead of trained immunity [2]. It has been shown that low to moderate doses of β-glucan, tri-DAP, and muramyl dipeptides are reported to induce trained immunity [33]. It also needs to be cautioned that the effects of trained immunity are likely associated with individual's genetic and epigenetic background. For example, nonspecific effects of infant BCG vaccination are reported to be heterogeneous, affected by multiple genetic and environmental factors including age, gender, interactions with other vaccines, and exposure to infectious pathogens at the time of BCG vaccination [34].

2.3 Mediators of trained immunity

It has been reported that pre-administration of pro-inflammatory innate cytokines [interleukin-1 (IL-1), tumor necrosis factor-α (TNF-α), and IL-6] provided protection against a variety of microbes [35]. Among the cytokines administered, IL-1 showed superior effects over TNF-α or IL-6 [35]. In BCG-vaccinated individuals, increase in production of these innate cytokines by monocytes in response to other microbes, other than BCG, was also found; this effect was again the most dependent on IL-1β [32]. IL-1β has also been reported to be crucial in the induction of trained immunity in NK cells [36]. On the other hand, in individuals with chronic mucocutaneous candidiasis, STAT-1-mediated type II interferon (IFN) induction was found to be crucial for induction of trained immunity [37]. The role of type II IFN (IFN-γ) in animal models was also reported by Kaufmann et al. [16]. However, in humans, innate immunity-associated protection (trained immunity) has been mainly implicated with IL-1β and other IL-1 families [38].

As detailed in the previous section, a metabolic shift from oxidative phosphorylation to aerobic glycolysis through the Hypoxia inducible factor-1α (HIF-1α) pathway downstream to mTOR is crucial for the development of trained immunity, since inhibition of this pathway is abolished induction of trained immunity [21]. Namely, in HIF-1α knockout mice, trained immunity was not induced [21]. IL-1β is known to be a direct target of HIF-1α [39], having a HIF-1α binding site in the promoter region of IL-1β gene [40]. It is now thus proposed that HIF-1α-induced

IL-1β also plays a role in epigenetic changes, through histone modifications [35]. Alternatively, IL-1β has been shown to upregulate HIF-1α [41].

Given the role of IL-1β in trained immunity, excessive, dysregulated production of IL-1β is likely to cause maladapted trained immunity and resultant pathogenic consequences. This may be observed in patients with autoinflammatory syndromes associated with gene mutation that lead to overproduction of IL-1β, including cryopyrin associated periodic syndrome (CAPS) [38, 42]. On the other hand, impaired induction of trained immunity can also cause detrimental effects. It was reported that patients with chronic mucocutaneous candidiasis exhibit impaired induction of STAT-1-dependent, trained immunity in response to β-glucan [37].

The above-described metabolic shift is not limited to glucose metabolism. Changes in glutamine and cholesterol metabolism have also shown to be crucial in trained immunity [24]. Consequently, it is thought that increased cholesterol content also plays a role in the development of trained immunity. Interestingly, increased levels of oxidized low-density lipoprotein (OxLDL) caused by dysregulated cholesterol metabolism are found to induce trained immunity in human monocytes [10]. Such a finding indicates a pathogenic role for maladapted trained immunity in atherosclerosis, since monocyte and macrophage cells are known to play a major role in plaque formation in vascular endothelium, a major histologic change in atherosclerosis [10].

2.4 Tolerance in innate immunity

As detailed in the previous section, trained immunity causes a metabolic shift from oxidative phosphorylation (OXPHOS) to glycolysis, rendering macrophage and monocyte lineage cells to classically activated cells or M1 phenotype; these cells exhibit impaired OXPHOS and anabolic repurposing of the tricarboxylic acid (TCA) cycle [43, 44]. In contrast, alternatively activated or the M2 phenotype of macrophages and monocytes has balanced processes of OXPHOS and TCA cycle activation; enhanced glycolytic generation of pyruvate fuels the TCA cycle, paralleling the induction of OXPHOS [44]. Trained macrophages via ß-glucan exposure are shown to reveal M1 phenotype [21]. Generation of M1 vs. M2 phenotypes of macrophages indicates the importance of regulating innate immune responses for prevention of excessive, potentially harmful inflammatory responses. In addition to generation of M2 phenotype, hypo-responsiveness of innate immunity has been described as endotoxin tolerance and compensatory anti-inflammatory response syndrome (CARS) [45]. Such regulatory mechanisms also have lasting effects, as observed in trained immunity.

Endotoxin tolerance in innate immunity was first shown in rodent models of sepsis. Namely, survival from sepsis is associated with diminished or absence of responses to LPS, an endotoxin [46]. Subsequently, it was shown that previous exposure to a sublethal dose of LPS led to resistance to a lethal dose of LPS in rodents [46]. Endotoxin tolerance is thought to be a result of innate immune memory with lasting immune hypo-responsiveness, even to non-LPS stimulants [47]. Phenotypic changes of tolerant innate immune cells are characterized with less production of inflammatory cytokines (TNF-α, IL-12, IL-6) and increase in production of counter-regulatory cytokines (IL-10 and TGF-β) upon stimulation [48, 49]. CARS was recognized as a clinical syndrome which is thought to represent a phase of immune "exhaustion," following initial potent immune activation, known as systemic inflammatory response syndrome (SIRS) [50]. Peripheral blood monocytes and neutrophils from CARS patients are reported to reveal similar phenotype to endotoxin-tolerant cells observed in rodent models [45, 49]. Recent research revealed that persistent effects of endotoxin tolerance and CARS are mediated by lncRNAs as well as microRNAs (miRNAs).

LPS activates TLR4 which leads to the activation of the myeloid differentiation factor 88 (MyD88)-mediated pathway and the TIR-domain-containing adaptor-inducing interferon-β (TRIF) pathway [45]. The molecular signature of endotoxin tolerance involves downregulation of TLR4, decreased recruitment of MyD88 or TRIF to TLR4, decreased activation of IL-1 receptor-associated kinase (IRAK)1 and IRAK4, diminished nuclear factor κ chain of B-cell (NF-κB) signaling, as well as upregulation of negative regulatory molecules including SH2 domain-containing inositol phosphatase 1 (SHIP1) [51].

2.5 Regulators of innate immune tolerance

Recent research revealed a role of miRNAs in the regulation of endotoxin tolerance. Specifically, miR-155 and miR-146α have been shown to regulate endotoxin tolerance [52]. MiR-146α reduces TLR signaling, by targeting IRAK1 and TRAF6, key components of TLR signaling pathway [53]. In contrast, miR-155 is reported to inhibit expression of SHIP1 and SOCS1, negative regulators of TLR signaling, prohibiting or attenuating tolerance induction by endotoxin [54, 55]. Several other miRNAs are also implicated with regulation of endotoxin intolerance [45]. It was shown recently that miR-221/miR-222 regulates functional reprogramming of macrophages during LPS-induced tolerization [47]. miR-221/miR-222 targets brahma-regulated gene 1 (Brg1), rendering transcriptional silencing of a subset of inflammatory genes that depend on SWI/SNF and STAT-mediated chromatin remodeling [47].

Recent research also revealed a role of lncRNAs in endotoxin tolerance; lncRNAs exert transcriptional, posttranscriptional, and translational regulation of gene expression [56–58]. Multiple lncRNAs are reported to regulate target molecules of TLR4 signaling pathways. LPS-responsive lncRNAs Mirt2, THRIl, MALTAT1, NKILA, lincRNA-21, and SeT have been reported to suppress expression of pro-inflammatory mediators including TNF-α [45]. For example, Mirt2 is reported to inhibit TRAF6 ubiquitination, leading to a decrease in TNF-α production [59]. However, at this time, relationships between actions of miRNAs and lncRNA in innate immune tolerance are not well understood. Other soluble mediators such as cytokines (IL-1β, IL-10, TGF-β, and TNF-α) are also reported to induce cross-tolerance or cytokine-mediated tolerance, causing a signaling cascade similar to that observed in TLR signaling [60]. In contrast, interferons (IFN-γ, α2-IFN, etc.) are known to abrogate endotoxin tolerance [61, 62]. Again these soluble mediators exert their actions on endotoxin tolerance via modulation of intracellular lncRNAs [45].

This type of innate immune memory (tolerance) is thought to be important in maintaining brain homeostasis, and impaired tolerance of innate immunity has been suspected in chronic neurodegenerative conditions such as Alzheimer's disease [9]. Aging is associated with an increased load of gram-negative bacteria in the GI tract and mouth mucosa, resulting in an increase in endotoxin levels in the blood and the brain [62]. However, aging individuals tolerate higher LPS levels in the brain through developing endotoxin tolerance [63].

3. Role of innate immunity in the nervous system

It is known that innate immunity does exist in the brain, playing a crucial role in brain morphogenesis and homeostasis. The major innate immune cells in the central nervous system (CNS) are microglial cells which are endogenously generated in the brain, but they can also be developed from bone marrow-derived monocytes, which are called BM-derived microglial cells (BMDM) [64, 65]. BMDM-induced inflammation has been implicated in neuropsychiatric conditions [64, 65]. It has also been reported that peripherally derived macrophages modulate microglial function after CNS injury; in this case, they are reported to exert anti-inflammatory effects [66]. Other innate immune cells in the CNS such as astrocytes are also known to exert important physiological roles [9, 67].

3.1 Trained immunity in the CNS

Inflammation in the periphery can prompt immune responses in the brain [1, 4]. Given the effects of trained immunity (activation vs. tolerance) in rodent models and humans, the development of maladapted innate immune memory in the CNS is expected to result in undesired, hazardous effects to the brain. However, reports concerning the effects of trained immunity and/or innate immune tolerance in the brain have been limited. Nevertheless, it was shown that microglial cells isolated from adult rats that were exposed to *E. coli* during the newborn period had increased expression of IL-1β mRNA [68]. The rats exposed to *E. coli* as newborns were also found to have impaired memory when they were challenged with a low dose of LPS, which was blocked by minocycline [2]. In experiments employing microglial cells obtained from sheep fetuses whose mother was given LPS intravenously, these fetal microglial cells were shown to have metabolic and epigenetic modulation, as has been reported in trained immunity [69].

Independent of the studies concerning trained immunity in the brain, persistent effects of maternal immune activation (MIA) on fetuses have been extensively studied, as one of the best studied rodent models of ASD [70]. In this model, sterile inflammation in pregnant rodents was induced with the use of PAMPs such as LPS, poly I:C, resulting in impaired neuropsychiatric symptoms in offspring in their adult years [70]. That is, offspring of MIA mothers have been shown to suffer from persistent behavioral symptoms and cognitive deficits frequently seen in ASD subjects later in life [70]. In addition, MIA also causes persistent alteration of adaptive immunity [71]. However, in this model, it is not yet well understood how innate immune memory (most likely trained immunity in this model) plays a role in a MIA model, causing persistent behavioral changes and impaired cognitive development. Children exposed to stressful events during the fetal and newborn period have also been reported to have higher levels of pro-inflammatory cytokines and neurodevelopmental impairment than control children [2]. Given the research findings in molecular mechanisms of trained immunity described in the previous section, there is a possibility that maladapted trained immunity contributes to the onset and progress of some neuropsychiatric disorders.

3.2 Innate tolerance in the brain

Tolerized innate immunity in the brain is thought to be crucial for limiting excessive inflammatory responses during brain tissue repair that involves phagocytosis of apoptotic cells and damaged tissue debris by tolerant phagocytes [72]. In rodent models, disruption of this pathway leads to neuroinflammation and subsequent neuronal damage [73]. An important regulator of this pathway is the triggering receptor expressed on myeloid cells 2 (TREM-2), which is expressed on microglial cells [74]. Blockade of TREM-2 was shown to exacerbate experimental autoimmune encephalitis (EAE), a rodent model of multiple sclerosis (MS) [75]. Apolipoprotein E (ApoE) which is a TREM-2 ligand was shown to have a role in maintaining tolerized phenotype of phagocytic cells [74]. This interaction was found to be impaired in patients with Alzheimer's disease [9]. In animal models of Alzheimer's disease treated with trained immunity vs. tolerance inducing stimuli, it was reported that long-term modulation of brain immune responses were observed, and the authors attributed this prolonged effects on innate immune memory to reprogramming of microglial cells [4].

3.3 mTOR-related pathology in neuropsychiatric disorders

In the previous section describing molecular pathways associated with trained immunity, the importance of mTOR signaling has been repeatedly shown. One thing we learned from the research on trained immunity is that multiple lineage cells reveal metabolic and epigenetic reprogramming in the process of innate immune memory, which, in animal models, can also be applied to microglial cells [4]. Interestingly, brain dysfunction caused by dysregulated mTOR signaling has been implicated in several neuropsychiatric disorders. In the next paragraph, we summarize mTOR-related brain dysfunctions and proposed mechanisms.

One of the expected consequences of excessive mTOR signaling caused by trained immunity is the impairment of lysosomal degradation of intracellular components, since mTOR activation inhibits autophagy via inhibition of the early steps of autophagosome biogenesis [76, 77]. Autophagy is a key physiological cellular function that clears intracellular molecules and thought to be developed to adjust the state of nutrient depletion [76, 77]. However, this is also an important mechanism to remove misfolded proteins that naturally occur in living cells [22]. In addition to degradation of misfolded proteins, autophagy also degrades altered subcellular organelles, such as the mitochondria [22]. Prolonged dysfunction in autophagy can lead to detrimental effects and is implicated in the pathogenesis of multiple neuropsychiatric conditions including dementia, movement disorders, seizures, brain ischemia, ASD, affective disorder, and schizophrenia [78–82]. In rodent models of depression, tuberous sclerosis, and ASD, rapamycin (sirolimus), a representative mTOR inhibitor, has been shown to attenuate social interactions and reverse behavioral effects on their neuropsychiatric symptoms [83–86]. Thus metabolic and epigenetic changes caused by trained immunity may have profound effects through altered levels of autophagy, as a result of metabolic and epigenetic reprograming, as detailed in the previous section.

3.4 ASD and a possible role of trained immunity

In this section, we discuss a possible role for trained immunity in the onset and progress of ASD. As a clinician, the author observed that an apparent strong immune stimulus altered the responses to subsequent immune stimuli in some, but not all ASD children and these ASD children also exhibit fluctuating neuropsychiatric symptoms, following microbial infection [87, 88]. As discussed in the previous section, in the MIA model of ASD, prolonged effects of MIA on the offspring brain can be explained through a concept of trained immunity occurring to the fetus at the time of sterile immune activation in the mother. This may have also happened in ASD subjects as described above. However, it should be noted that ASD is a behaviorally defined syndrome, diagnosed on the basis of behavioral symptoms, except for a minority of ASD cases that have well-defined gene mutations [89]. Therefore, based on the author's clinical experience, it is likely that trained immunity plays a role in a subset of ASD subjects for whom neuroinflammation is associated in their ASD pathogenesis.

In ASD patients, just like in other neuropsychiatric conditions, a role of inflammation has been long suspected, and more and more evidence has been accumulating [90–92]. In the research of innate immune abnormalities in ASD children, we have also found evidence of dysregulated innate immune responses, shifting to pro-inflammatory responses in a subset of ASD subjects [88, 93, 94]. We also experienced that these ASD subjects suffer from various comorbid medical conditions involving the gastrointestinal (GI) tract and other organs [87]. Retrospectively, our findings may be reflecting maladapted innate immunity as a form of trained immunity in such ASD subjects; these ASD subjects may fall into an ASD subset which we have called inflammatory autism, mimicking the rodent ASD model of MIA [93]. Our previous findings that may indicate altered innate immune memory in such ASD patients are as follows:

- In some but not all the ASD subjects, we found significant changes in innate immune abnormalities which are best reflected in changes in IL-1β/IL-10 ratios produced by purified peripheral blood monocytes (PBMo) [88, 93]. Namely, some patients reveal high ratios of IL-1β/IL-10, while others showed low ratios, and these rations can change from time to time, depending on their exposure to immune insults [93].

- ASD subjects who revealed high and/or low IL-1β/IL-10 ratios also revealed fluctuating behavioral symptoms following immune insults [94]. Parents of these subjects often describe more severe, prolonged illnesses and frequent respiratory infection following microbial infection [87]. They also seem to reveal significant changes in their behavioral symptoms and cognitive activity with immune stimuli not associated with microbial infection; these ASD children may exhibit worsening neuropsychiatric symptoms, following flare-ups of aeroallergen allergy, delayed-type food allergy, and adverse reactions to medications including vaccinations [87, 94].

- ASD subjects who revealed high and/or low IL-1β/IL-10 ratios also revealed changes in production of inflammatory monocyte cytokines including TNF-α and IL-6 [93, 95].

- PBMo from ASD subjects who revealed altered IL-1β/IL-10 ratios also revealed changes in miRNA expression by PBMo, as compared to cells obtained from neurotypical, non-ASD controls [93].

- We also studied changes in mitochondrial respiration in peripheral blood mononuclear cells (PBMCs) obtained from ASD subjects and non-ASD controls. Our results revealed evidence of altered mitochondrial respiration in association with changes in IL-1β/IL-10 ratios by PBMo in ASD subjects [95].

- In recent studies, we also found changes in miRNA in sera of ASD subjects, when tested by high-throughput deep sequencing. Again, changes in serum miRNA levels are closely associated with changes in IL-1β/IL-10 ratios by PBMo, production of monocyte cytokines (TNF-β, IL-6, IL-10, CCL2 mostly), along with parameters of mitochondrial respiration (manuscript submitted for publication). Interestingly, in ASD subjects, miRNA levels are mostly decreased, as compared to non-ASD controls (submitted for publication). Targeted genes by miRNAs that are altered in serum levels in ASD subjects with high or low IL-1β/IL-10 ratios are associated with pathways involved in innate immune responses, including the mTOR signaling pathway (unpublished observation).

The above-described findings may be best explained by altered innate immune responses associated with innate immune memory (trained immunity vs. tolerance). So, if this is the case, for these ASD subjects, can clinical features that indicate an alternation of innate immune memory be detected? The author is a pediatric immunologist and, as indicated before, as stated previously, observes exacerbation of neuropsychiatric symptoms, following immune insults. Herein, a representative ASD case, in which trained immunity may be associated with the onset and progression of ASD, is presented.

3.5 Case presentation

A 10-year-old female child presented to the pediatric allergy/immunology clinic at our institution secondary to fluctuating behavioral symptoms. Fluctuation of behavioral symptoms often occurred, following microbial infection.

The patient was born at 41 weeks of gestation via cesarean section due to breech presentation, following an uneventful pregnancy. The patient was developing typically until 24 months of age and then suffered from significant developmental regression. Prior to the onset of the developmental regression, parents took the patient to South Asia to visit other family members and friends. During this visit, the patient suffered an insect bite which was complicated by a secondary bacterial skin infection. When treated with oral antibiotics abroad, the patient developed generalized hives and severe GI symptoms (nausea, vomiting, diarrhea, and bloating): the patient then became intolerant to multiple foods. After returning to the United States, the patient was given multiple vaccinations including live vaccines to catch up the vaccination schedule. All these vaccines were given while the patient was still suffering from GI symptoms and an active skin infection. Within several days after vaccinations (multiple vaccines given all together), noticeable loss of cognitive and motor skills became apparent in the patient. The patient was eventually diagnosed with ASD around 2.5 years of age.

Eventually, the patient's GI symptoms subsided, but this subject never regained the cognitive skills that this patient had once acquired prior to the onset of developmental regression. Prior to advancing to pre-kindergarten, the patient was given booster doses of vaccines which were well tolerated. However, after starting pre-kindergarten, the patient started getting sick frequently with upper respiratory infections, which often evolved into ear infection. The patient missed many days of school, since the patient suffered a prolonged course of illness and more severe symptoms, as compared to peers. While the patient presented with symptoms of upper respiratory infection, this patient's behavioral symptoms continue to fluctuate, most evident in worsening of obsessive compulsive behaviors and frequency of "rage" episodes. Worsening behavioral symptoms would always follow immune insults, worse in a convalescence stage. Avoidance of sick contacts by placing the patient in home schooling attenuated the fluctuating behavioral symptoms. At 7–8 years of age, the fluctuating behavioral symptoms seen were mainly associated with teething. After the completion of teething, behavioral symptoms became more stable. However, the patient stopped growing, falling under the first percentile of the growth curve in height and weight. An exhausting workup for primary mitochondrial diseases, endocrine diseases, primary immunodeficiency with known gene mutations, and congenital metabolic and genetic diseases was unrevealing. However, video electric encephalogram revealed a focal epileptic activity. Family history is negative for neuropsychiatric, genetic, autoimmune, immune, and metabolic diseases.

In the case presented above, did neuroinflammation caused by maladapted trained immunity have a role in her clinical features? It is hard to prove, but it may be speculated that the initial stressful events that occurred abroad shaped trained immunity in this patient, and the subsequent multiple unrelated immune stimuli

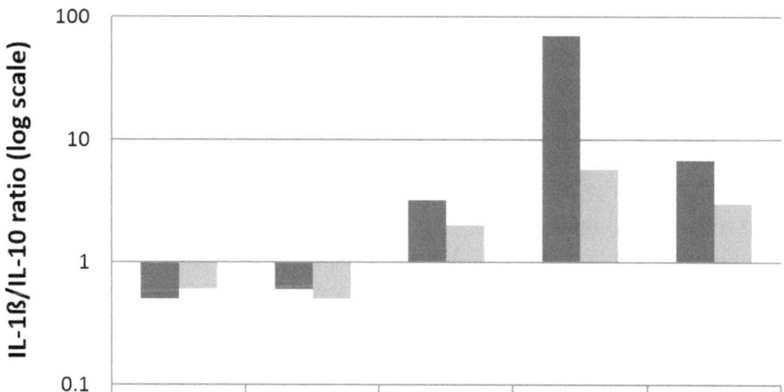

	medium	LPS	zymosan	CL097	ß-glucan
Patient	0.5	0.6	3.2	69.7	6.8
Control	0.6	0.5	2	5.7	3

Figure 1.
IL-1β/IL-10 ratios produced by purified peripheral blood monocytes in response to medium only (no stimulus), LPS (TLR4 agonist), zymosan (TLR2/TLR6 agonist), CL097 (TLR7/TLR8 agonist), and β-glucan in the presented case (patient) and control cells from a non-ASD neurotypical subject. IL-1β/IL-10 ratios are shown in a log scale.

may have caused prolonged maladapted trained immunity, leading to persistent neuroinflammation and impairment of cognitive activity, as observed in the MIA models of ASD. Interestingly, changes in GI conditions, such as changes in microbiome, have been implicated with neuropsychiatric diseases, triggering maladapted trained immunity [96]. It is also reported that trained innate immunity can be induced in human monocytes by cow's milk [97]. Thus her severe GI symptoms and subsequent intolerance to multiple foods may be associated with excessive trained immunity in the gut of this patient.

3.6 Evidence of impaired trained immunity

As summarized in the previous section, we have found that IL-1β/IL-10 ratios produced by PBMo are altered in some ASD subjects in association with fluctuating behavioral symptoms [94]. Thus if innate immune memory (trained immunity) is associated with her above-described remarkable clinical symptoms, we may also find altered IL-1β/IL-10 ratios, as an indicator of altered innate immune responses.

Thus we assessed IL-1β/IL-10 ratios produced by PBMo in response to a panel of innate immune stimuli, including β-glucan, as reported previously [95]. As shown in **Figure 1**, the presented case revealed increase in IL-1β/IL-10 ratios in response to zymosan, CL097, and β-glucan. High IL-1β/IL-10 ratio in response to CL097, an agonist of TLR7/TLR8, was especially striking. We also observed increase in production of TNF-α and IL-6 and decrease in the production of IL-10, as well. Given these findings, it is possible that maladapted trained immunity may have caused excessive inflammatory responses to various innate immune stimuli, which then led to developmental regression and fluctuating behavioral symptoms in this presented case.

4. Conclusions

Our deepening knowledge of innate immune memory (trained immunity vs. tolerance) has shed light on the understanding of nonspecific effects of microbial infection and other immune stimuli, which have been implicated in the onset and progress of various neuropsychiatric diseases. Recent research indicates a possibility for a role of maladapted innate immune memory in various neuropsychiatric conditions. The finding of innate immune memory is especially exciting in the field of neuroimmunology, since we now likely have better tools for addressing the long-suspected role of immune-mediated inflammation that is not associated with specific pathogens or environmental factors, in various neuropsychiatric conditions. The concept of innate immune memory will be especially important in addressing insults to the brain during the early years of CNS development, and the resultant lasting intellectual disabilities, as seen in MIA models [70]. More importantly, an improved understanding of the role of innate immune memory (trained immunity vs. tolerance) in pathogenic neuroinflammation can lead to novel therapeutic measures that are desperately needed for the treatment of neuropsychiatric diseases.

Acknowledgements

The part of the study presented in this manuscript was funded from the Jonty Foundation, St. Paul, MN. The author is thankful for the critical review of this manuscript by Dr. L. Huguenin.

Conflict of interest

The author has nothing to declare.

Abbreviations

Ab	antibody
Ag	antigen
APC	Ag-presenting cells
ApoE	apolipoprotein E
ASD	autism spectrum disorders
BCG	Bacillus Calmette-Guérin
Brg1	brahma-regulated gene 1
BMDM	cell, bone marrow-derived microglial cell
CAPS	cryopyrin-associated periodic syndrome
CARS	compensatory anti-inflammatory response syndrome
CNS	central nervous system
DAMPs	damage-associated molecular patterns
EAE	experimental autoimmune encephalitis
GI	gastrointestinal
HIF-1α	hypoxia inducible factor-1α
IFN	interferon
IL	interleukin
IPLs	immune gene-priming lncRNAs
lncRNAs	long noncoding RNAs
IRAK	interleukin-1 receptor-associated kinase
LPS	lipopolysaccharide
MIA	maternal immune activation
MS	multiple sclerosis
MyD88	myeloid differentiation factor 88
mTOR	mammalian target of rapamycin

NF-κB	nuclear factor of κ chain of B cells
NK	natural killer
OxLDL	oxidized low-density lipoprotein
OXPHOS	oxidative phosphorylation
PAMPs	pathogen-associated molecular patterns
PBMCs	peripheral blood mononuclear cells
PBMo	peripheral blood monocytes
SHIP1	SH2 domain-containing inositol phosphatase 1
SPUH	Saint Peter's University Hospital
TCA	tricarboxylic acid
TLR	Toll-like receptor
TNF	tumor necrosis factor
TREM-2	triggering receptor expressed on myeloid cells 2
TRIF	TIR-domain-containing adaptor-inducing interferon-ß
TSC	tuberous sclerosis complex
UMLILO	upstream master lncRNAs of the inflammatory chemokine locus

Author details

Harumi Jyonouchi[1,2]

1 Department of Pediatrics, Rutgers-RWJ, Saint Peter's University Hospital (SPUH), New Brunswick, NJ, United States

2 Rutgers-Robert Wood Johnson Medical School, New Brunswick, NJ, United States

*Address all correspondence to: hjyonouchi@saintpetersuh.com

References

[1] Perry VH, Cunningham C, Holmes C. Systemic infections and inflammation affect chronic neurodegeneration. Nature Reviews. Immunology. 2007;**7**(2):161-167

[2] Salam AP, Borsini A, Zunszain PA. Trained innate immunity: A salient factor in the pathogenesis of neuroimmune psychiatric disorders. Molecular Psychiatry. 2018;**23**(2):170-176

[3] van der Heijden C, Noz MP, Joosten LAB, Netea MG, Riksen NP, Keating ST. Epigenetics and trained immunity. Antioxidants & Redox Signaling. 2018;**29**(11):1023-1040

[4] Wendeln AC, Degenhardt K, Kaurani L, Gertig M, Ulas T, Jain G, et al. Innate immune memory in the brain shapes neurological disease hallmarks. Nature. 2018;**556**(7701):332-338

[5] Krakauer T. Inflammasomes, autophagy, and cell death: The trinity of innate host Defense against intracellular Bacteria. Mediators of Inflammation. 2019;**2019**:2471215

[6] Peng Y, van Wersch R, Zhang Y. Convergent and divergent signaling in PAMP-triggered immunity and effector-triggered immunity. Molecular Plant-Microbe Interactions. 2018;**31**(4):403-409

[7] McCoy KD, Ronchi F, Geuking MB. Host-microbiota interactions and adaptive immunity. Immunological Reviews. 2017;**279**(1):63-69

[8] Farber DL, Netea MG, Radbruch A, Rajewsky K, Zinkernagel RM. Immunological memory: Lessons from the past and a look to the future. Nature Reviews. Immunology. 2016;**16**(2):124-128

[9] Sfera A, Gradini R, Cummings M, Diaz E, Price AI, Osorio C. Rusty microglia: Trainers of innate immunity in Alzheimer's disease. Frontiers in Neurology. 2018;**9**:1062

[10] Sohrabi Y, Lagache SMM, Schnack L, Godfrey R, Kahles F, Bruemmer D, et al. mTOR-dependent oxidative stress regulates oxLDL-induced trained innate immunity in human monocytes. Frontiers in Immunology. 2018;**9**:3155

[11] Benn CS, Netea MG, Selin LK, Aaby P. A small jab—A big effect: Nonspecific immunomodulation by vaccines. Trends in Immunology. 2013;**34**(9):431-439

[12] Aaby P, Roth A, Ravn H, Napirna BM, Rodrigues A, Lisse IM, et al. Randomized trial of BCG vaccination at birth to low-birth-weight children: Beneficial nonspecific effects in the neonatal period? The Journal of Infectious Diseases. 2011;**204**(2):245-252

[13] Kleinnijenhuis J, Quintin J, Preijers F, Joosten LA, Ifrim DC, Saeed S, et al. Bacille Calmette-Guerin induces NOD2-dependent nonspecific protection from reinfection via epigenetic reprogramming of monocytes. Proceedings of the National Academy of Sciences of the United States of America. 2012;**109**(43):17537-17542

[14] Sun JC, Beilke JN, Lanier LL. Adaptive immune features of natural killer cells. Nature. 2009;**457**(7229):557-561

[15] Mitroulis I, Ruppova K, Wang B, Chen LS, Grzybek M, Grinenko T, et al. Modulation of Myelopoiesis progenitors is an integral component of trained immunity. Cell. 2018;**172**(1-2):147-161 e12

[16] Kaufmann E, Sanz J, Dunn JL, Khan N, Mendonca LE, Pacis A, et al. BCG educates hematopoietic stem cells to generate protective innate immunity

against tuberculosis. Cell. 2018; **172**(1-2):176-190 e19

[17] Suliman S, Geldenhuys H, Johnson JL, Hughes JE, Smit E, Murphy M, et al. Bacillus Calmette-Guerin (BCG) revaccination of adults with latent mycobacterium tuberculosis infection induces long-lived BCG-reactive NK cell responses. Journal of Immunology. 2016;**197**(4):1100-1110

[18] Jensen KJ, Larsen N, Biering-Sorensen S, Andersen A, Eriksen HB, Monteiro I, et al. Heterologous immunological effects of early BCG vaccination in low-birth-weight infants in Guinea-Bissau: A randomized-controlled trial. The Journal of Infectious Diseases. 2015;**211**(6):956-967

[19] Netea MG, Joosten LA, Latz E, Mills KH, Natoli G, Stunnenberg HG, et al. Trained immunity: A program of innate immune memory in health and disease. Science. 2016;**352**(6284):aaf1098

[20] Reimer-Michalski EM, Conrath U. Innate immune memory in plants. Seminars in Immunology. 2016;**28**(4):319-327

[21] Cheng SC, Quintin J, Cramer RA, Shepardson KM, Saeed S, Kumar V, et al. mTOR- and HIF-1alpha-mediated aerobic glycolysis as metabolic basis for trained immunity. Science. 2014;**345**(6204):1250684

[22] Ryskalin L, Limanaqi F, Frati A, Busceti CL, Fornai F. mTOR-related brain dysfunctions in neuropsychiatric disorders. International Journal of Molecular Sciences. 2018;**19**(8):1-29

[23] Arts RJ, Joosten LA, Netea MG. Immunometabolic circuits in trained immunity. Seminars in Immunology. 2016;**28**(5):425-430

[24] Arts RJ, Novakovic B, Ter Horst R, Carvalho A, Bekkering S, Lachmandas E, et al. Glutaminolysis and fumarate accumulation integrate immunometabolic and epigenetic programs in trained immunity. Cell Metabolism. 2016;**24**(6):807-819

[25] Saeed S, Quintin J, Kerstens HH, Rao NA, Aghajanirefah A, Matarese F, et al. Epigenetic programming of monocyte-to-macrophage differentiation and trained innate immunity. Science. 2014;**345**(6204):1251086

[26] Novakovic B, Habibi E, Wang SY, Arts RJW, Davar R, Megchelenbrink W, et al. Beta-glucan reverses the epigenetic state of LPS-induced immunological tolerance. Cell. 2016;**167**(5):1354-1368 e14

[27] Su X, Wellen KE, Rabinowitz JD. Metabolic control of methylation and acetylation. Current Opinion in Chemical Biology. 2016;**30**:52-60

[28] Shin Y, Chang YC, Lee DSW, Berry J, Sanders DW, Ronceray P, et al. Liquid nuclear condensates mechanically sense and restructure the genome. Cell. 2018;**175**(6):1481-1491 e13

[29] Cho WK, Spille JH, Hecht M, Lee C, Li C, Grube V, et al. Mediator and RNA polymerase II clusters associate in transcription-dependent condensates. Science. 2018;**361**(6400):412-415

[30] Fok ET, Davignon L, Fanucchi S, Mhlanga MM. The lncRNA connection between cellular metabolism and epigenetics in trained immunity. Frontiers in Immunology. 2018;**9**:3184

[31] Fanucchi S, Fok ET, Dalla E, Shibayama Y, Borner K, Chang EY, et al. Immune genes are primed for robust transcription by proximal long noncoding RNAs located in nuclear compartments. Nature Genetics. 2019;**51**(1):138-150

[32] Nissen TN, Birk NM, Blok BA, Arts RJW, Andersen A, Kjaergaard J, et al. Bacillus Calmette-Guerin vaccination at birth and in vitro cytokine responses to non-specific stimulation. A randomized clinical trial. European Journal of Clinical Microbiology & Infectious Diseases. 2018;**37**(1):29-41

[33] Ifrim DC, Quintin J, Joosten LA, Jacobs C, Jansen T, Jacobs L, et al. Trained immunity or tolerance: Opposing functional programs induced in human monocytes after engagement of various pattern recognition receptors. Clinical and Vaccine Immunology. 2014;**21**(4):534-545

[34] Butkeviciute E, Jones CE, Smith SG. Heterologous effects of infant BCG vaccination: Potential mechanisms of immunity. Future Microbiology. 2018;**13**:1193-1208

[35] Netea MG, van der Meer JW. Trained immunity: An ancient way of remembering. Cell Host & Microbe. 2017;**21**(3):297-300

[36] Kleinnijenhuis J, Quintin J, Preijers F, Joosten LA, Jacobs C, Xavier RJ, et al. BCG-induced trained immunity in NK cells: Role for non-specific protection to infection. Clinical Immunology. 2014;**155**(2):213-219

[37] Ifrim DC, Quintin J, Meerstein-Kessel L, Plantinga TS, Joosten LA, van der Meer JW, et al. Defective trained immunity in patients with STAT-1-dependent chronic mucocutaneaous candidiasis. Clinical and Experimental Immunology. 2015;**181**(3):434-440

[38] Moorlag S, Roring RJ, Joosten LAB, Netea MG. The role of the interleukin-1 family in trained immunity. Immunological Reviews. 2018;**281**(1):28-39

[39] Tannahill GM, Curtis AM, Adamik J, Palsson-McDermott EM, McGettrick AF, Goel G, et al. Succinate is an inflammatory signal that induces IL-1beta through HIF-1alpha. Nature. 2013;**496**(7444):238-242

[40] Zhang W, Petrovic JM, Callaghan D, Jones A, Cui H, Howlett C, et al. Evidence that hypoxia-inducible factor-1 (HIF-1) mediates transcriptional activation of interleukin-1beta (IL-1beta) in astrocyte cultures. Journal of Neuroimmunology. 2006;**174**(1-2):63-73

[41] Jung YJ, Isaacs JS, Lee S, Trepel J, Neckers L. IL-1beta-mediated up-regulation of HIF-1alpha via an NFkappaB/COX-2 pathway identifies HIF-1 as a critical link between inflammation and oncogenesis. The FASEB Journal. 2003;**17**(14):2115-2117

[42] Keddie S, Parker T, Lachmann HJ, Ginsberg L. Cryopyrin-associated periodic fever syndrome and the nervous system. Current Treatment Options in Neurology. 2018;**20**(10):43

[43] O'Neill LA, Kishton RJ, Rathmell J. A guide to immunometabolism for immunologists. Nature Reviews. Immunology. 2016;**16**(9):553-565

[44] Van den Bossche J, Baardman J, de Winther MP. Metabolic characterization of polarized M1 and M2 bone marrow-derived macrophages using real-time extracellular flux analysis. Journal of Visualized Experiments. 2015;(105):53424

[45] Vergadi E, Vaporidi K, Tsatsanis C. Regulation of endotoxin tolerance and compensatory anti-inflammatory response syndrome by non-coding RNAs. Frontiers in Immunology. 2018;**9**:2705

[46] Zhang R, Miller RG, Gascon R, Champion S, Katz J, Lancero M, et al. Circulating endotoxin and systemic immune activation in sporadic amyotrophic lateral sclerosis (sALS). Journal of Neuroimmunology. 2009;**206**(1-2):121-124

[47] Seeley JJ, Baker RG, Mohamed G, Bruns T, Hayden MS, Deshmukh SD, et al. Induction of innate immune memory via microRNA targeting of chromatin remodelling factors. Nature. 2018;**559**(7712):114-119

[48] del Fresno C, Garcia-Rio F, Gomez-Pina V, Soares-Schanoski A, Fernandez-Ruiz I, Jurado T, et al. Potent phagocytic activity with impaired antigen presentation identifying lipopolysaccharide-tolerant human monocytes: Demonstration in isolated monocytes from cystic fibrosis patients. Journal of Immunology. 2009;**182**(10):6494-6507

[49] Nahid MA, Satoh M, Chan EK. MicroRNA in TLR signaling and endotoxin tolerance. Cellular & Molecular Immunology. 2011;**8**(5):388-403

[50] Ward NS, Casserly B, Ayala A. The compensatory anti-inflammatory response syndrome (CARS) in critically ill patients. Clinics in Chest Medicine. 2008;**29**(4):617-625, viii

[51] Biswas SK, Lopez-Collazo E. Endotoxin tolerance: New mechanisms, molecules and clinical significance. Trends in Immunology. 2009;**30**(10):475-487

[52] Doxaki C, Kampranis SC, Eliopoulos AG, Spilianakis C, Tsatsanis C. Coordinated regulation of miR-155 and miR-146a genes during induction of endotoxin tolerance in macrophages. Journal of Immunology. 2015;**195**(12):5750-5761

[53] Vergadi E, Vaporidi K, Theodorakis EE, Doxaki C, Lagoudaki E, Ieronymaki E, et al. Akt2 deficiency protects from acute lung injury via alternative macrophage activation and miR-146a induction in mice. Journal of Immunology. 2014;**192**(1):394-406

[54] O'Connell RM, Chaudhuri AA, Rao DS, Baltimore D. Inositol phosphatase SHIP1 is a primary target of miR-155. Proceedings of the National Academy of Sciences of the United States of America. 2009;**106**(17):7113-7118

[55] Sun X, Sun J, Shao X, Feng J, Yan J, Qin Y. Inhibition of microRNA-155 modulates endotoxin tolerance by upregulating suppressor of cytokine signaling 1 in microglia. Experimental and Therapeutic Medicine. 2018;**15**(6):4709-4716

[56] Magistri M, Faghihi MA, St Laurent G 3rd, Wahlestedt C. Regulation of chromatin structure by long noncoding RNAs: Focus on natural antisense transcripts. Trends in Genetics. 2012;**28**(8):389-396

[57] Ulitsky I, Bartel DP. lincRNAs: Genomics, evolution, and mechanisms. Cell. 2013;**154**(1):26-46

[58] Murphy MB, Medvedev AE. Long noncoding RNAs as regulators of toll-like receptor signaling and innate immunity. Journal of Leukocyte Biology. 2016;**99**(6):839-850

[59] Du M, Yuan L, Tan X, Huang D, Wang X, Zheng Z, et al. The LPS-inducible lncRNA Mirt2 is a negative regulator of inflammation. Nature Communications. 2017;**8**(1):2049

[60] Mia S, Warnecke A, Zhang XM, Malmstrom V, Harris RA. An optimized protocol for human M2 macrophages using M-CSF and IL-4/IL-10/TGF-beta yields a dominant immunosuppressive phenotype. Scandinavian Journal of Immunology. 2014;**79**(5):305-314

[61] Chen J, Ivashkiv LB. IFN-gamma abrogates endotoxin tolerance by facilitating Toll-like receptor-induced chromatin remodeling. Proceedings of the National Academy of Sciences of the United States of America. 2010;**107**(45):19438-19443

[62] Shi L, Song L, Maurer K, Sharp J, Zhang Z, Sullivan KE. Endotoxin tolerance in monocytes can be mitigated by alpha2-interferon. Journal of Leukocyte Biology. 2015;**98**(4):651-659

[63] Zhan X, Stamova B, Sharp FR. Lipopolysaccharide associates with amyloid plaques, Neurons and Oligodendrocytes in Alzheimer's Disease Brain: A Review. Frontiers in Aging Neuroscience. 2018;**10**:42

[64] Wlodarczyk A, Cedile O, Jensen KN, Jasson A, Mony JT, Khorooshi R, et al. Pathologic and protective roles for microglial subsets and bone marrow- and blood-derived myeloid cells in central nervous system inflammation. Frontiers in Immunology. 2015;**6**:463

[65] Ataka K, Asakawa A, Nagaishi K, Kaimoto K, Sawada A, Hayakawa Y, et al. Bone marrow-derived microglia infiltrate into the paraventricular nucleus of chronic psychological stress-loaded mice. PLoS One. 2013;**8**(11):e81744

[66] Greenhalgh AD, Zarruk JG, Healy LM, Baskar Jesudasan SJ, Jhelum P, Salmon CK, et al. Peripherally derived macrophages modulate microglial function to reduce inflammation after CNS injury. PLoS Biology. 2018;**16**(10):e2005264

[67] Tchessalova D, Posillico CK, Tronson NC. Neuroimmune activation drives multiple brain states. Frontiers in Systems Neuroscience. 2018;**12**:39

[68] Williamson LL, Sholar PW, Mistry RS, Smith SH, Bilbo SD. Microglia and memory: Modulation by early-life infection. The Journal of Neuroscience. 2011;**31**(43):15511-15521

[69] Cao M, Cortes M, Moore CS, Leong SY, Durosier LD, Burns P, et al. Fetal microglial phenotype in vitro carries memory of prior in vivo exposure to inflammation. Frontiers in Cellular Neuroscience. 2015;**9**:294

[70] Careaga M, Murai T, Bauman MD. Maternal immune activation and autism Spectrum disorder: From rodents to nonhuman and human primates. Biological Psychiatry. 2017;**81**(5):391-401

[71] Rose DR, Careaga M, Van de Water J, McAllister K, Bauman MD, Ashwood P. Long-term altered immune responses following fetal priming in a non-human primate model of maternal immune activation. Brain, Behavior, and Immunity. 2017;**63**:60-70

[72] Fu R, Shen Q, Xu P, Luo JJ, Tang Y. Phagocytosis of microglia in the central nervous system diseases. Molecular Neurobiology. 2014;**49**(3):1422-1434

[73] Neher JJ, Neniskyte U, Brown GC. Primary phagocytosis of neurons by inflamed microglia: Potential roles in neurodegeneration. Frontiers in Pharmacology. 2012;**3**:27

[74] Painter MM, Atagi Y, Liu CC, Rademakers R, Xu H, Fryer JD, et al. TREM2 in CNS homeostasis and neurodegenerative disease. Molecular Neurodegeneration. 2015;**10**:43

[75] Ohrfelt A, Axelsson M, Malmestrom C, Novakova L, Heslegrave A, Blennow K, et al. Soluble TREM-2 in cerebrospinal fluid from patients with multiple sclerosis treated with natalizumab or mitoxantrone. Multiple Sclerosis. 2016;**22**(12):1587-1595

[76] Hosokawa N, Hara T, Kaizuka T, Kishi C, Takamura A, Miura Y, et al. Nutrient-dependent mTORC1 association with the ULK1-Atg13-FIP200 complex required for autophagy. Molecular Biology of the Cell. 2009;**20**(7):1981-1991

[77] Jung CH, Jun CB, Ro SH, Kim YM, Otto NM, Cao J, et al. ULK-Atg13-FIP200 complexes mediate mTOR signaling to the autophagy machinery. Molecular Biology of the Cell. 2009;**20**(7):1992-2003

[78] Heras-Sandoval D, Perez-Rojas JM, Hernandez-Damian J, Pedraza-Chaverri J. The role of PI3K/AKT/mTOR pathway in the modulation of autophagy and the clearance of protein aggregates in neurodegeneration. Cellular Signalling. 2014;**26**(12):2694-2701

[79] Carloni S, Buonocore G, Balduini W. Protective role of autophagy in neonatal hypoxia-ischemia induced brain injury. Neurobiology of Disease. 2008;**32**(3):329-339

[80] Giorgi FS, Biagioni F, Lenzi P, Frati A, Fornai F. The role of autophagy in epileptogenesis and in epilepsy-induced neuronal alterations. Journal of Neural Transmission (Vienna). 2015;**122**(6):849-862

[81] Tramutola A, Triplett JC, Di Domenico F, Niedowicz DM, Murphy MP, Coccia R, et al. Alteration of mTOR signaling occurs early in the progression of Alzheimer disease (AD): Analysis of brain from subjects with pre-clinical AD, amnestic mild cognitive impairment and late-stage AD. Journal of Neurochemistry. 2015;**133**(5):739-749

[82] Ryskalin L, Lazzeri G, Flaibani M, Biagioni F, Gambardella S, Frati A, et al. mTOR-dependent cell proliferation in the brain. BioMed Research International. 2017;**2017**:7082696

[83] Cleary C, Linde JA, Hiscock KM, Hadas I, Belmaker RH, Agam G, et al. Antidepressive-like effects of rapamycin in animal models: Implications for mTOR inhibition as a new target for treatment of affective disorders. Brain Research Bulletin. 2008;**76**(5):469-473

[84] Sato A, Kasai S, Kobayashi T, Takamatsu Y, Hino O, Ikeda K, et al. Rapamycin reverses impaired social interaction in mouse models of tuberous sclerosis complex. Nature Communications. 2012;**3**:1292

[85] Tang G, Gudsnuk K, Kuo SH, Cotrina ML, Rosoklija G, Sosunov A, et al. Loss of mTOR-dependent macroautophagy causes autistic-like synaptic pruning deficits. Neuron. 2014;**83**(5):1131-1143

[86] Kara NZ, Flaisher-Grinberg S, Anderson GW, Agam G, Einat H. Mood-stabilizing effects of rapamycin and its analog temsirolimus: Relevance to autophagy. Behavioural Pharmacology. 2018;**29**(4):379-384

[87] Jyonouchi H, Geng L, Streck DL, Toruner GA. Children with autism spectrum disorders (ASD) who exhibit chronic gastrointestinal (GI) symptoms and marked fluctuation of behavioral symptoms exhibit distinct innate immune abnormalities and transcriptional profiles of peripheral blood (PB) monocytes. Journal of Neuroimmunology. 2011;**238**(1-2):73-80

[88] Jyonouchi H, Geng L, Davidow AL. Cytokine profiles by peripheral blood monocytes are associated with changes in behavioral symptoms following immune insults in a subset of ASD subjects: An inflammatory subtype? Journal of Neuroinflammation. 2014;**11**:187

[89] Turner TN, Eichler EE. The role of De novo noncoding regulatory mutations in neurodevelopmental disorders. Trends in Neurosciences. 2019;**42**(2):115-127

[90] Cristiano C, Lama A, Lembo F, Mollica MP, Calignano A, Mattace RG. Interplay between peripheral and central inflammation in autism Spectrum disorders: Possible nutritional

and therapeutic strategies. Frontiers in Physiology. 2018;**9**:184

[91] Jiang NM, Cowan M, Moonah SN, Petri WA Jr. The impact of systemic inflammation on neurodevelopment. Trends in Molecular Medicine. 2018;**24**(9):794-804

[92] Siniscalco D, Schultz S, Brigida AL, Antonucci N. Inflammation and neuro-immune dysregulations in autism spectrum disorders. Pharmaceuticals (Basel). 4 June 2018;**11**(2):56

[93] Jyonouchi H, Geng L, Streck DL, Dermody JJ, Toruner GA. MicroRNA expression changes in association with changes in interleukin-1ss/interleukin10 ratios produced by monocytes in autism spectrum disorders: Their association with neuropsychiatric symptoms and comorbid conditions (observational study). Journal of Neuroinflammation. 2017;**14**(1):229

[94] Jyonouchi H, Geng L, Buyske S. Interleukin-1β/Interleukin10 ratio produced by monocytes as a biomarker of neuroinflammation in autism. Journal of Clinical & Cellular Immunology. 2017;**8**(3):1000503

[95] Jyonouchi H, Geng L, Rose S, Bennuri SC, Frye RE. Variations in mitochondrial respiration differ in IL-1ß/IL-10 ratio based subgroups in autism spectrum disorders. Frontiers in Psychiatry. 2019;**10**(Feburary):1-15

[96] Rudzki L, Szulc A. "Immune gate" of psychopathology-the role of gut derived immune activation in major psychiatric disorders. Frontiers in Psychiatry. 2018;**9**:205

[97] van Splunter M, van Osch TLJ, Brugman S, Savelkoul HFJ, Joosten LAB, Netea MG, et al. Induction of trained innate immunity in human monocytes by bovine milk and milk-derived immunoglobulin G. Nutrients. 27 Sept 2018;**10**(10)

Chapter 6

Cytokines in Scar Glial Formation after an Acute and Chronic Spinal Cord Injury

Roxana Rodrígez-Barrera, Adrián Flores-Romero, Julián García-Sánchez, Lisset Karina Navarro-Torres, Marcela Garibay-López and Elisa García-Vences

Abstract

The inflammatory response after a spinal cord injury (SCI) is a secondary mechanism of damage, this involves alterations at the local and systemic level, and it is mediated by cytokine participation that takes part actively. The excessive inflammatory response causes an autoreactive response that targets against components of the nervous tissue; this response lengthens the inflammatory process initiated during the acute phase. The participation of immune cells in acute phases is characterized by the arrival of neutrophils, macrophages, and microglia, as well as T lymphocytes, which express their peaks on different days post-injury (1st, 3rd, and 11th respectively). The chronic phase of the injury begins 14 days after it occurred, reaching its highest point at 60 days, and can still be detected the following 180 days. One of the outcomes of the inflammatory process and cytokine synthesis is the generation of glial scar. In this chapter, we will review the different cytokine mechanisms involved in the formation of glial scar in acute and chronic phases, as well as the modulating treatments of glial scar.

Keywords: spinal cord injury, immune cells, scar glial, modulating treatments

1. Introduction

Spinal cord injury (SCI) causes catastrophic damaged to patients, and the incidence is getting higher each year. Most of them are occasioned by physical trauma from sports injuries, car accidents, falls, and more [1, 2]. This life-changing neurological condition also comes with socioeconomic implications for patients and their caregivers, besides the functional and sensitive consequences that are largely determined by the level and completeness of the injury [1, 3].

After SCI, the acute and focal inflammation triggers a multicellular and multifunctional complex response which induces resident and infiltrating cells to form the glial scar (GS) at the site of the lesion [3]. The GS is a complicated phenomenon which has been considered as one of the main causes of limited regenerative capacity by inhibiting axonal regeneration and preventing functional recovery [4]. It has been proven that the GS creates both a physical barrier for neural repair as well as a chemical inhibition by the secretion of inhibitory extracellular matrix

 IntechOpen

molecules [5]. At the present time, finding an effective treatment has shown to be challenging due to the lack of complete understanding about the multifactorial pathophysiology of SCI. Current medical treatment is confined to surgical procedures and anti-inflammatory drugs which aim to reduce the damage caused by the continuous inflammatory reactions and therefore increase the locomotor recovery. More importantly, recent studies have demonstrated that the GS can be both favorable or prejudicial depending on the evolution time of the SCI, being able to participate in tissue repair and functional recovery during the acute phase but later on establishing a recovery plateau due to the inhibition of axonal regeneration during the chronic phase [6, 7]. Therefore, in recent years there has been an increasing interest in developing new therapies that can modulate the immunological responses involved in the GS formation. Although there are many drugs that have been identified as potential treatments for SCI, there is currently no therapy that can effectively restore the neural function that is lost during this pathology. The purpose of this chapter is to describe the importance of cytokines in the immunological processes of GS formation as well novel therapies that could serve as potential treatments of SCI.

2. Inflammation in traumatic central nervous system

Disorders in the homeostasis of the central nervous system (CNS) just as infection, trauma, ischemia, neurodegenerative diseases, and disturbances in general induce the beginning of neuroinflammatory responses that can be considered to consist principally of innate immune mechanisms [8, 9].

Inflammation is the way the human body acts in response to situations such as injury and infection. This mechanism involves several processes of the somatosensory, autonomic, immune, and vascular systems and more [10].

The immune and nervous system are capable of regulating physiological homeostasis and defending against infection and injury through inflammation. Both systems have improved many features for the recognition of alterations in the changing microenvironment to facilitate the protective responses. Although cells in each system (neurons and immune cells) have many differences, they can interact and communicate together to make a functional cooperation for the integral homeostasis [11].

Neuroinflammation is a localized inflammation in both CNS and peripheral nervous system (PNS), despite being distinct from the inflammation in peripheral tissues. There is also upregulation of several pro-inflammatory cytokines like IL-1β, IL-6, TNFα, and chemokines that affect the integrity of the blood brain barrier (BBB) resulting in local and systemic immune responses [9, 10, 12].

During these events, neural control plays an important role due to the fact that many immune molecules are detected by sensory neurons, which lead the system to generate immunoregulatory responses [13]. In general, the nervous system integrates biological functions to restore homeostatic function with the use of neurotransmitters and other regulatory molecules [11].

On the other hand, there are other cell types involved in the response, such as microglial and astrocytes. Microglial cells are made to deal with the harmful effects involving the activation of astrocytes, which are capable of modulating the activity of other immunocompetent cells in the site of the injury and also have an active role in the synaptic elimination, regeneration, cell elongation, and repair [8, 14, 15]. Many studies have reported that astrocytes participate in axonal regeneration by providing growth substrates and guidance structures [16]. They are also required in CNS repair, especially in the acute phase after injury but not in the chronic phase, reducing GS formation and exacerbating the magnitude and duration of inflammatory activation [17, 18].

Nevertheless, it has been confirmed that inflammatory mechanisms contribute both to cell damage and tissue remodeling [12]. They are involved in reactive plasticity modulation of neuronal populations in different types of brain injuries, as well as in microglial cells and astrocytes, since they can activate and promote recovery and repair the neural circuits [8].

Owing to the events explained before, neuroinflammation is taken into account as an important manipulable aspect of injury in animal and human studies to develop immunomodulatory therapies because it can be detrimental or beneficial; however, it is necessary to understand the processes in a better way [19].

2.1 Inflammatory response after spinal cord injury

In the SCI there are two phases of the pathology: the primary one which consists in the initial accident-induced damage that can result from the compression of the spinal cord (SC), stretching of the nervous tissue, or disruption of local blood supply and the secondary one, which is characterized by the damage caused by inflammation and other biological mechanisms. These events can start at the moment of the injury and go on for days, and even weeks, after the event [2, 20] (see **Figure 1**).

Inflammatory microenvironment after SCI involves activated microglia, astrocytes, and infiltrating macrophages that play a role in the development of the secondary injury, and it is the major target to combat SCI [1, 2]. Then, this environment following SCI is mediated by the activation of microglia and astrocytes and infiltrating macrophages that greatly contribute to the progression of secondary injury that is a compilation of complex events derived from the initial trauma. Some of the mechanisms in its pathogenesis includes neurodegeneration, gliosis, and apoptosis in nearby intact neural tissues [21–25]. Effective restraint of secondary injury is essential to minimize neurodegeneration and to improve significantly functional recovery [1].

It is important to know that a traumatic injury in the CNS begins with the disruption of the BBB and blood-spinal barrier (BSB), followed by the arrival of several cells and molecules of the immune system with the possibility of aggravating the situation, affecting subsequent events such as repair and regeneration [26].

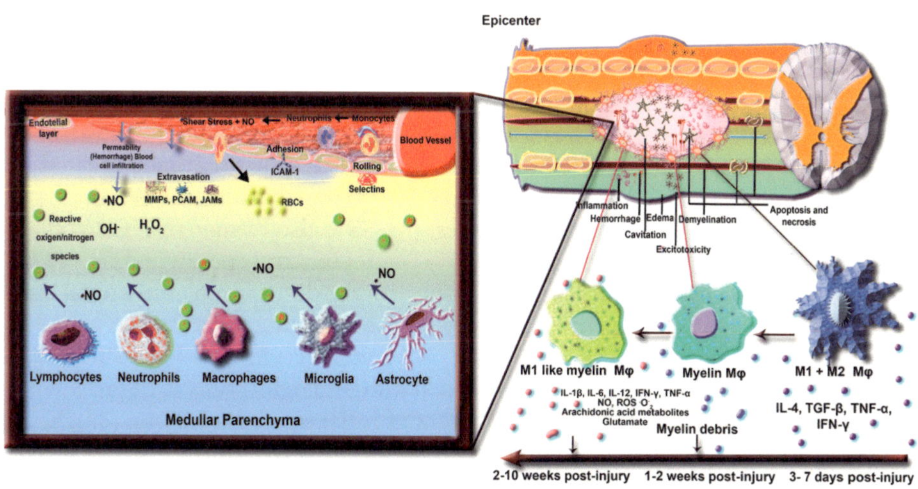

Figure 1.
Inflammatory response after spinal cord injury.

The inflammatory microenvironment after SCI involves activated microglia, astrocytes, and infiltrating macrophages that play a role in the development of the secondary injury, and it is the major target to combat SCI.

It has been demonstrated that there is a multiphasic response during the inflammation processes after SCI and a huge interaction between central and peripheral cellular and soluble components, which are influenced by some factors like patient age, sex, mechanism and degree of injury, therapeutic interventions, and genetic variability [9, 26].

In a study made by Beck et al. [26], they established two types of cellular inflammation phases. The early phase includes principally the infiltration of neutrophils which are polymorphonuclear leucocytes (PMNs), macrophages, and microglia, which inhibit recovery of the brain and SCI after the traumatic event. It was discovered that neutrophils are peaking 1 day of post-injury, macrophages/microglia 7 days of post-injury, and T cells 9 days of post-injury. The late phase was detected after 14 days of post-injury with its peak after 60 days of post-injury, and it also remained detectable throughout the 180 days of post-injury for all three cell types mentioned before. Moreover, the inhibition of the C5a-mediated inflammation after 14 days of injury reduced the locomotor recovery and myelination of the SC in the damaged site, suggesting that the late phase involves a restorative function [26].

Following this line, after the damage to the microvasculature, the presence of progressive edema and proapoptotic signaling begins. All of these events promote thrombosis and microvessel spasms causing hypoxia. Therefore, a relevant aspect to mention is that astrocytes are the first to act in the injury site, contributing to the formation of the GS, as well as preventing neurons to grow and heal [1, 20]. In the same way, neutrophils and macrophages are recruited from the periphery to the injured area, and, together with reactive astrocytes, microglia/macrophages will also contribute to the formation of a regeneration-inhibiting GS [27].

3. Cytokines and acute glial scar formation after spinal cord injury

In the early phase, the formation of GS has a protective function, isolating potentially dangerous molecules of the rest of healthy tissue and controlling the spread of damage [28]. Accordingly with this, GS is considered as a mechanism of protection, developed by the organism against injuries that affect the CNS.

Overall, the GS is composed of two parts, the fibrotic and the glial. The fibrotic scar occupies the core of the injury with deposits of collagen matrix and is mainly composed of invading fibroblasts derived from meningeal and perivascular cells [29, 30]. The GS occupies the peripheral zone of the lesion and is composed mainly of astrocytes due to its evolution from a dynamic process known as reactive astrogliosis [30]. The glial limiting membrane is a specialized structure that is located close to the outer layer of the fibrotic scar and marks the division between these two parts [31]. Besides fibroblasts and astrocytes, the GS is also formed by NG2 + oligodendrocyte precursor cells (OPCs), microglia, pericytes, and ependymal cells [32].

A phenomenon that occurs simultaneously with the destruction of neuronal components is the activation of an inflammatory response characterized at first moment by the release of chemokines by endothelial cells and microglia cells [33]. These chemokines induce the migration of peripheral immunological cells to the affected tissue and promote the posterior establishment of inflammatory response [34]. The migration of macrophages and lymphocytes besides the activation of microglial cells is joined to the deficiency to control an inflammatory process in the CNS, thereby contributing with a destructive immunological response [35]. Both resident and infiltrating cells contribute to the GS formation, and the main characteristic of the inflammatory response at this time point is the sustained production of free radicals due to the continuous synthesis of pro-inflammatory cytokines like TNF-α, IL-1β, INF-γ, and enzymes that activate glial cells or disrupt the BSB [34, 36]. Moreover, activated macrophages produce and secrete matrix metalloproteinases (MMPs) to furtherly disrupt the BSB and increase vascular permeability [37].

Another phenomenon observed is the activation of inflammasome. The damage of cell membranes permits the release of molecules of ATP and the efflux of K^+, stimulating the activation of the inflammasome and inducing the production of the pro-inflammatory cytokines, IL-1β and IL-18, which these cytokines have been related with neurodegenerative process [38]. Furthermore, studies in vitro have shown the direct relation of IL-1β with the overexpression of glial fibrillary acidic protein (GFAP) on astrocytes; for this reason the activation of inflammasome is a key factor involved in the formation and maturation of GS [39].

Astrocyte is a specific cell residing only in the CNS that maintains the homeostasis, conforms the BBB, and keeps the concentration of ions and neurotransmitters to regulate the activity in neuronal synapsis [40].

Reactive astrocytes (RAs) possess surface receptors for different cytokines just like the cells of the immune system, making them a target for products derived from the inflammatory environment. The pro-inflammatory cytokines induce the upregulation of inflammatory genes on astrocytes and the posterior secretion of various chemokines, including CXCL1, CCL2, CCL3, CCL4, and CXCL12, and cytokines like IL-1, transforming growth factor β (TGFβ), TNF-α, and INFγ. For example, INFγ interacts to modulate several facets of the gliotic response, and such interactions with growth factors may be important in creating the biochemical and physical properties of the GS; for this reason this cytokine is responsible for failed neuronal regeneration after SCI [41]. In this way, the astrocytes can contribute with the presence of a constant inflammatory response, affecting themselves and influencing other cell populations related with the formation of GS [41, 42].

Fibroblasts secrete extracellular matrix (ECM) components which include the chondroitin sulfate proteoglycans (CSPGs) family (neurocan, versican, brevican, phosphocan, and NG2) which is mainly secreted by astrocytes as well as fibronectin, collagen, and laminin which are produced by fibroblasts [36, 43]. Altogether, they contribute to the formation of GS and participate in developing its characteristic impermeability and the expression of molecules that impede the anatomical and functional restoration after the lesion [32, 44]. Fibroblasts also possess cytokine receptors on their surface which respond to high concentrations of pro-inflammatory cytokines and stimulate the activation of enzymatic machinery synthesized ECM protein. The inhibition of pericytes and fibroblasts by the application of different therapeutical strategies reduce the size and consolidation of GS, showing the importance of these types of cells in the GS formation. Of great importance is the presence of specific receptor for TGFβ on fibroblasts; the stimulation with this molecule facilitates the synthesis and release of collagen type IV [45].

Generally speaking, after an injury to the CNS, there is a sequential phenotypic change in astrocytes called reactive astrogliosis, where naïve astrocytes (NAs) are transformed to RAs which eventually become scar-forming astrocytes (SAs) that can inhibit axonal regeneration and functional recovery [7]. Astrocytes are the most abundant glial cells within the CNS, and although they are not part of the immune system, they play a crucial role in the pathophysiology of the GS formation [36].

Furthermore, RAs substantially upregulate their expression of GFAP, intermediate filaments, nestin, and vimentin and mobilize to the center of the injury to form a mesh-like structure of interlaced filamentous structures [46]. A high concentration of pro-inflammatory cytokines induces upregulation of GFAP on astrocytes and the development of hypertrophic prolongation. A certain study showed that genetically modified mice with deficiency of vimentin and GFAP produced a less dense GS which frequently conducted to constant bleeding, suggesting that vimentin and GFAP are part of the main cytoskeletal intermediate filaments that form the GS [47].

This astrocytic migration secludes inflammatory cells from the surrounding intact tissues and minimize the extension of secondary damage after CNS injury leading to tissue repair and functional improvement during the acute phase of GS formation [7, 48, 49]. In addition, the hemorrhagic flow into the CNS due to the rupture of the BBB exposes scar-forming cells to factors in plasma such as fibrinogen which has been proven to induce the expression of CSPGs in astrocytes through TGFβ/Smad2 signaling pathway [50].

Besides, in the acute phase of the GS formation, the overexpression of CSPGs (neurocan, versican, brevican, phosphocan, and NG2+) plays a beneficial role by modulating the inflammatory activity of resident microglia as well as the infiltration of monocytes through the CD44 receptor [51].

Moreover, to the featured RAs, the GS formation also requires the activation of ependymal cells, NG2+-expressing glia (including OPCs), meningeal- and vascular-derived fibroblasts, pericytes, and macrophages surrounding the injury area [52]. More importantly, some of these previously mentioned cells have the capacity to switch their phenotypes and become RAs to furtherly contribute in the GS formation [53]. Furthermore, there are several molecular mechanisms that contribute to the formation of the GS such as the upregulation of bone morphogenetic proteins (BMPs), MMPs, epidermal growth factor receptor (EGFR), eph/ephrins, TGFβ, and signal transducer and activator of transcription and interleukin (STAT/IL) family (STAT3) [30, 54, 55]. The upregulation of BMP-4 has shown to promote astrocyte differentiation and to inhibit the production of oligodendrocytes and neurons [56]. In addition, the MMP family is involved in the ECM remodeling, and therefore, they are structurally and temporally involved in the GS formation [57]. The limitation of the extent of the GS was seen with the suppression of MMP-2 in mice, and MMP-9 has proven to be involved in the augmented migration of RAs to the injury site, therefore facilitating GS formation [58]. EGFR is upregulated in astrocytes following damage to the SC, leading to the activation of the Rheb-mToR signaling pathway which induces astrocytes to migrate and suffer hypertrophy to furtherly form the GS [59]. Moreover, EGFR ligands, such as transforming growth factor-alpha (TGF-alpha) and EGF, contribute to the formation of the GS by inducing astrocytes to secrete CSGPs [60]. In addition, TGFβ expression is upregulated immediately after SCI. It promotes the formation of the GS by simultaneously stimulating monocyte and lymphocyte activity as well as inducing the production and deposition of new ECM proteins (collagen, fibronectin, and proteoglycans) [61, 62]. The manipulation of TGFβ signaling in the injured CNS modulates the formation of the fibrotic scar in the lesion site. The administration of TGFβ1 to the injured CNS increases the deposition of ECMs in the lesion site [63, 64], while antibodies to TGFβ1 and TGFβ2 and the endogenous TGFβ inhibitor decorin, a small leucine-rich CSPG, conversely reduce the size of GS [64], which proposes the involvement of TGFβs in the formation of GS. In addition, RAs release TNF-α to inhibit oligodendrocyte progenitor cell (OPG) survival and prevent them from differentiating into mature oligodendrocytes, suggesting a mechanism for the failure of remyelination after SCI [65].

4. Modifications of the glial scar

RAs have been traditionally considered to be a unidirectional and irreversible process; however, recent studies have proven to inhibit its progress and even revert the astrocyte's phenotype according to environmental cues [48, 49]. Over the past few years, there has been an increasing interest in modulating the GS formation; nevertheless, there has been a wide spectrum of results mainly due to the fact that the GS has many components and there are many different types of therapy strategies. Inclusive, recent studies have shown that the attenuation of RAs to prevent GS formation has resulted in a worse outcome in SCI and limited functional recovery [6, 7]. In transgenic mice where STAT3 selectively suppressed RAs showed reduced migration to the lesion epicenter, leading to an extensive area of injury with uncontrolled inflammatory cell filtration and limited functional recovery [7]. Another study showed a pronounced reduction of glial scarring in animals with conditional knockdown of STAT3, suggesting that this molecule is one of the most important factors involved in the formation of the GS [62]. A wide spectrum of molecules such as type I and II interferons and cytokines, growth factors including EGF, platelet-derived growth factor, IL-6, leukemia inhibitory factor, and ciliary neurotrophic factor (CNF) is able to activate STAT3 in order to cause variations in RAs and elicit GS formation [66, 67]. Similarly, a recent study used HSV1tk/GCV (a suicide system gene) to selectively kill proliferating RAs in SCI to avoid GS formation, resulting in a widespread infiltration of inflammatory cells and continuous involvement of healthy tissue surrounding the epicenter of the lesion as well as decreased neuronal survival and decreased locomotor recovery [5]. These findings furtherly support that reactive astrogliosis in acute-subacute phases plays beneficial roles in acute wound healing, remodeling processes, and isolating the injury to prevent the spread of cytotoxic molecules and inflammatory cells into the surrounding tissue [4, 5].

Even though GS formation in acute phases has proven to have beneficial effects, its evolution and persistence in chronic stages of the injury have shown to become a strong inhibitor for SC regeneration [3]. Therefore, there has been some attempts in regulating the chronic phase of the GS to improve axonal outgrowth.

5. Cytokines and chronic glial scar formation after spinal cord injury

Through the years, it became clear that both the scar tissue and the immune system play important beneficial roles in axonal regeneration and healing of the CNS [68].

As mentioned before, SCI results in the disruption of the BBB, and the BSB increased inflammatory reactions such as the activation of the microglia and the production of various cytokines and augmented the activation of TGFβ and Smad2 signaling pathways [49]. The inflammatory microenvironment presented after the insult continues in most of the cases until the chronic phase [34].

Acute GS formation restricts inflammation and preserves neural tissue [28, 46, 69]. Nonetheless, at the chronic phase (>14 days after the injury in mice), RAs progressively transform into SAs that form astrocytic scars which compose the main impediment for axonal regeneration and functional recovery in the chronic phase of SCI [70, 71]. It has been suggested that after inflammation has resolved, chronic GS is expendable and detrimental because it continually prevents axon regeneration [6]. For this reason, it is necessary in chronic phases to inhibit, modulate, or remove the mature GS.

Certain factors present during the acute formation of the glial scar are also active during its chronic formation. The genetic suppression of BMPR1b (a subtype of the BMP type 1 receptor) resulted in the weakening of the GS in chronic stages of SCI, suggesting that BMPs play an important role in the acute formation of the GS as well as in its stabilization through the chronic stage [72].

Although the expression of CSPGs during acute glial scar formation participates in reducing the damage extent, the prolonged exposure of CSPGs is prejudicial for functional recovery for they are well-known to be the main inhibitors of GS axonal regeneration, sprouting, and remyelination during the chronic phase of SCI [3, 73]. The posterior formation of the GS traps on its core of GS, where they reside contributing with the chronic presence of an inflammatory response. The continual synthesis of pro-inflammatory cytokines like TNFα, IL1β, and INFγ promotes the aggregation of new elements and the modification of GS [74].

This is the main reason why astrogliosis may cause both beneficial and detrimental effects depending on its dynamic features and on its time course [50, 69]. Cytokines behave in a similar way. In the early stages of GS formation, pro-inflammatory cytokines such as TNF-α, IL-1β, and INF-γ help by recruiting and activating microglial cells, astrocytes, and other peripheral immunological cells to the injury site to prevent the extension of the injury [33, 35]. However, other acute-secreted cytokines such as IL-1β and IL-18 have been associated with neurodegenerative processes and activation of the inflammasome [37, 38]. On the other hand, during the chronic formation of the GS, cytokines contribute to impede axonal regeneration and functional recovery. Overall, cytokines may present both beneficial and detrimental effects depending on the stage of GS formation and depending on the process in which they are involved. Certain cytokines which are present in early stages and may present beneficial effects by increasing the production of CSPGs may become prejudicial as time progresses. Current pharmacological treatments

Cytokine	Scar glial effect	References
TGFβ	Induces the expression of CSPGs in astrocyte Facilitates the synthesis and release of collagen type IV in fibroblast Increases the deposition of ECMs in the lesion site Inhibits the generation of TGFβ1, TGFβ2, and SOX-9 and as a result there is a decreased deposition of CSPG	[60] [45] [63, 64] [28]
IL-1β	Increases overexpression of GFAP on astrocytes and maturation of GS	[39]
TNFα	RAs release TNF-α to inhibit OPCs survival and prevent them from differentiating into mature oligodendrocytes, suggesting a mechanism for the failure of remyelination after SCI Reduces the expression of GFAP through anti-inflammatory processes and helps to suppress reactive gliosis	[27]
INFγ	Promotes SG formation and modulates ECM which helps that the interactions with growth factors may be important in creating modification in the GS	[36]
IL-4 and IL13 IL-4 and IL-10	Transplantation with BMSCs was associated with significant increases in IL-4 and IL-13; these changes were associated with less scar tissue formation INDP in combination with scar removal and DPY reduces pro-inflammatory cytokines in chronic phase	[79] [78]
IL-10	Activates beneficial M2 macrophages which were found to regulate scar resolution	[18]

Table 1.
Main cytokines involve in scar glial formation.

depend on cytokines to establish their mechanism of action and should be focused to develop further pharmacological strategies.

The majority of SCI patients are those with lesions who may benefit insufficiently from therapeutic treatments designed for application in the chronic stage and focused on cytokines and other immunological processes. However, compared to treatments of acute experimental SCI, the efficacy of therapies promoting axonal regeneration seems impaired in chronic models. Therefore, GS formation can be improved if we combine treatments like stem cell transplants [75], iron chelators [76–78], and matrix biocompatible [16, 78]. **Table 1** summarizes some GS effects exerted by cytokines.

6. Modulate, inhibit, or remove glial scar as therapeutic tool

In this section we will review some modulating treatments of the GS. That should be able to counteract posttraumatic factors of inhibitory growth and promote axonal and tissue recovery.

6.1 Anti-inflammatory therapy

In contrast with the pro-inflammatory cytokines produced after the injury, the application of anti-inflammatory therapies like the treatment with doses of methylprednisolone (MP) after the injury avoids the formation of the GS. The application of MP is helpful to reduce the expression of GFAP and reduce the deposition of CSPG and avoid the formation of the GS [80].

Combination therapy using MP and tranilast after SCI in rats significantly reduced posttraumatic SC edema and neutrophil infiltration and improved functional recovery better than single individual therapies, and it also significantly reduced the amount of GFAP expression at the injury site [81].

Therapies that induce elevated concentrations of IL-10, a well-known anti-inflammatory cytokine, reduce in an important way the presence of CSPG on GS [82]. Astrocytes also express the transcription nuclear factor (NF-kB). The selective inhibition of NF-kB induces a better neurological outcome and a reduction in size of the GS. In addition, the interference of NF-kB induces the reduction of pro-inflammatory cytokines, chemokines, and secretion of CSPG [83].

Curcinum is a phytochemical compound that has an anti-inflammatory effect. This molecule inhibits pro-inflammatory cytokines (TNFα and IL-1β), which contribute to reduce the expression of GFAP through anti-inflammatory processes and help to suppress reactive gliosis [26]. Previous studies have also demonstrated to inhibit the generation of TGFβ1, TGFβ2, and SOX-9; as a result, there is a decreased deposition of CSPG, causing the inhibition of TGFβ and transcription factors. There is also evidence that curcinum reduces the amount of nestin and GFAP around de SCI, suggesting that it inhibits astrogliosis improving the microenvironment to SC repair [27].

In addition, rapamycin is an immunosuppressant that inhibits the mTor pathway selectively, and, it is considered neuroprotective because it increases the anti-inflammatory microenvironment and reduces locomotor impairment and damage in neural tissue. Other outcomes have shown that reduced infiltrations of macrophages and neutrophils at the SCI also reduce microglial activation and secretion of TNFβ; the amount of cells expressing GFAP inhibits proliferation of astrocytes and promotes angiogenesis and neuronal survival around the injury [2, 30].

Finally, TGFβ is involved in GS formation process, increasing the expression of neurocan, a CSPG that mediates GS formation and inhibits axon growth. Therefore, the use of antibodies against TGFβ1 and TGFβ2 is necessary; they mitigate the response of GFAP, causing the interruption of scar tissue and glial membrane formation that limit the edge of the injury. Astrocytes, OPCs, and NG2+ responses are diminished. This is possible by interrupting the Smad 3 signaling pathway in conjunction with TGFβ [2, 84].

In contrast, a study carried out by Kohta and colleagues showed that the inhibition of TGFβ1 with a neutralizing antibody resulted in the suppression of the GS formation resulting in a mild improvement of growth and/or preservation of axons in the injured GS caudal to the site of contusion [31]. Furthermore, rats treated with anti-TGFβ1 increased the activation of the microglia after injury, apparently providing a beneficial environment for the restoration and healing of the neural network [31].

The macrophages are immune cells with phagocytic capabilities. There are three subgroups, but now the focus goes towards M2 macrophages. When M2 macrophages infiltrate the SC, they may also secrete protective factors, such as the anti-inflammatory cytokine IL-10, and boosting the release of protective molecules would be advantageous. In addition, when MMP-2 is upregulated, it represents a beneficial effect for SCI recovery; however, when there is a deficiency in MMP-2-expression, an exacerbated lesion expansion, scar formation, vascular instability, and locomotor deficits are present [18, 85].

6.2 Cell therapies

6.2.1 Schwann cells

The SCs are the principal glia of the peripheral nervous system (PNS) [86], and in a SCI, they have shown to promote axonal regeneration through the formation of bridges across the injury. This bridge is a multicellular structure that crosses the lesion from the rostral to caudal part, providing an environment in which axons can grow and cover the GS to suppress axonal regeneration impediment [87, 88].

SC transplant provides a neuroprotective effect, preventing neural death by continuous inflammatory reaction caused by a SCI; moreover the neural peripheral grafts promote the expression of neurotrophins like BDNF and NGF, which is key for a successful regeneration as it delays the formation of the GS [89]. It is not advisable to transplant the SCs alone, because their regenerative capacity is limited by the secretion of myelin-associated and axonal growth inhibitors (CSPGs, semaphorins, and myelin-associated proteins) by the GS. Although many types of cells have been studied for transplantation, the SCs have always been considered as one of the best proposals for this treatment; however, they need to be co-transplanted with other molecules or cells such as OECs, MSCs, and NSCs, among other cells, in order to achieve its full therapeutic potential [88, 90].

6.2.2 Bone marrow mesenchymal stem cells

Bone marrow stem cells (BMSCs) are the most abundant cells in the bone marrow; they are hematopoietic and functional support cells [91]. The implantation of BMSCs has shown to have regenerative and immunomodulative properties that help to prevent the GS formation [91]. Furthermore, these cells are able to regulate CNTF-STAT3 signal transduction which reduces tissue scarring, inflammatory responses, and apoptosis [92]. Okuda et al. reported that BMSC sheets suppress the GS and provide a positive environment for axonal regeneration, causing changes in reactive astrocyte morphology [93]. Moreover, BMSCs can secrete different trophic factors (VEGF, BDNF, NGF, and hepatocyte growth factor) which increase positive results associated with BMSC transplantation [91, 94]. In addition, the transplant of BMSCs are associated with significant increases in IL-4 and IL-13; these changes were associated with less scar tissue formation [79].

With all that said, BMSCs possess many features that make them eligible for cell culture transplantation; however there are still many knowledge gaps that need to be studied, such as their survival rate when transplanted.

6.2.3 Olfactory ensheathing cells

Olfactory ensheathing cells (OECs) form the glial component of the primary olfactory system, and they reside both on CNS and PNS [87, 95]. Recent olfactory bulb (OB) transplants have shown to be able to infiltrate the scar tissue, through the environment of astrocytes thanks to their heparin profile [87]. They also provide a scaffold that promotes neuronal growth and angiogenesis and supply a bridge through the injury site that decreases the contusion area [96, 97]. OECs promote neural regeneration by promoting cell-to-cell interaction with sensorial axons and migrate ahead to the olfactory bulb, creating a favorable environment for axonal growth where cellular debris are phagocytized to increase restoration, neuroinflammation is modulated, neuroprotection is provided, and the expression of neurotrophic factors like BNDF, GDNF, NGF, and ECM molecules is augmented to provide a substrate for newly generated axons [98, 99]. These cells inhibits pro-inflammatory cytokines and induces the activation anti-inflammatory cytokines; they can activate neurotrophic factors.

Neurotrophic factors secreted by OECs are capable of inhibiting scar formation and promote axonal regeneration, implying that they also are neuroprotective. The receptors of each neurotrophic factor are NGF/p75, BDNF/Trκβ, GDNF/GFR-1, NTN/GFR-2, and NRG-1/ErbB [43]. Also OECs reduce the expression of GFAP by an earlier shorter immune response by astrocytes and microglia, due to the attenuation of NF-κβ, which is involved in RAs [100].

6.3 Chondroitinase ABC

The chondroitinase ABC (ChABC) is a bacterial enzyme that catalyzes the removal of the CSPG and therefore digests them. The administration of ChABC has demonstrated to inhibit CSPG and deactivate their glycosaminoglycan chains, which promotes a significant regeneration of axons, and M2 macrophage phenotype activation [2, 101].

Certain studies have used ChABC in SCI models in rats to enzymatically degrade CSPGs and therefore reduce its inhibitory functions in axonal regeneration [35, 36]. The results of these experiments showed a significant improvement in locomotor and proprioceptive functions, demonstrating that the degradation of CSPGs is a promising strategy to avoid its long-term prejudicial effects in chronic SCI [36, 37]. It is also reported that the combination of glial-derived neurotrophic factor (GDNF) and transplanted SCs causes a reduction in astrogliosis (GFAP and CSPG) and is also responsible for promoting axon regeneration after SCI [102]. Another combination therapy with ChABC, acidic fibroblast growth factor (aFGF), and peripheral nerve graft bridge supports axon regeneration and functional recovery after chronic SCI like so bladder physiology outcomes associated with an invasive repair strategy. CSPG are significantly downregulated by the astroglial NF-kB inhibition [83]. Taken all together, these studies demonstrate that the degradation of CSPGs is a promising strategy to avoid its long-term prejudicial effects in chronic SCI.

6.4 Iron chelators to inhibit collagen biosynthesis

Using iron chelators to inhibit collagen biosynthesis has been demonstrated to have beneficial effects by transient suppressing fibrous scarring in an acute SCI model [45, 77]. The iron chelation of α,α'-dipyridyl (DPY) has previously shown to decrease the collagen synthesis at a posttranscriptional level by inhibiting 4-prolyl hydroxylase, one of the key enzymes in collagen metabolism [103].

In a study with unilateral SC transection in adult and postnatal mice (14 days old) where DPY was applied at the injury site, it was observed that collagen type IV deposits and axons showed the expression of tyrosine hydroxylase and these axons extended through the site of injury by reinnervating the striatum [104]. Conversely, iron chelators suppress GS but do not degrade the existing scar, meaning that this treatment is not transferable to chronic SCI where a mature lesion scar is present, with a plethora of axon growth-inhibitory molecules attached [105, 106].

6.5 Surgical resections

Some studies have shown that the surgical removal of the GS promotes the development of axons in the injured portion of the SC, suggesting that axonal reconnection is feasible [16, 107]. Another study indicates that the use of surgical resection of the GS by itself does not offer positive results, because at the time of incising and removing the tissue healing, the same mechanisms that are activated during the acute phase are reactivated, generating a second lesion [75].

On the other hand, one study showed that careful surgical resection of the scar and filling cavity with biocompatible matrices promotes a functional improvement in a full-section model [16].

Therefore, the treatment of SCI can be improved if the behavior of the GS with the combination of transplants [75], iron chelators [45, 77], and matrices is biocompatible [16]. Furthermore, Rodriguez and colleagues explored whether INDP in combination with scar removal and DPY provided an appropriate microenvironment to promote neural restoration in chronic SCI. They found an increased activity in genes encoding for IL4, TGFβ, BDNF, IGF1, and GAP43, as well as a decreased activity in genes encoding for TNFα and IFNγ. Moreover, there was a significant increment in the number of serotonergic (5-HT-positive) and catecholaminergic (TH-positive) fibers at the caudal segment of the GS [78].

7. Conclusions

Cytokines are incredibly involved in GS formation during the acute and chronic phases of SCI, participating in either beneficial or detrimental effects. To achieve the best possible results, it is necessary to maintain the anti-inflammatory microenvironment for more extended periods of time in order to promote axonal regeneration, M2 phenotype macrophage activation, and secretion of neurotrophic factors that are capable of inhibiting the GS formation in the chronic phase. Several clinical trials have shown different therapeutic strategies to modulate the formation of GS. Although those experiments have had a significant therapeutic potential in patients with SCI, there are still enormous knowledge gaps which need further investigation in order to develop a potential cure for SCI.

Acknowledgements

We gratefully acknowledge the Universidad Anáhuac México for their support on this chapter.

Conflict of interest

The authors declare no conflict of interest.

Author details

Roxana Rodrígez-Barrera, Adrián Flores-Romero, Julián García-Sánchez, Lisset Karina Navarro-Torres, Marcela Garibay-López and Elisa García-Vences*
Centro de Investigación en Ciencias de la Salud (CICSA), Universidad Anáhuac México Campus Norte, Huixquilucan, Estado de México, Mexico

*Address all correspondence to: edna.garcia@anahuac.mx

References

[1] Sun X, Jones ZB, Chen XM, Zhou L, So KF, Ren Y. Multiple organ dysfunction and systemic inflammation after spinal cord injury: A complex relationship. Journal of Neuroinflammation. 2016;**13**(1):260

[2] Ahmed A, Patil AA, Agrawal DK. Immunobiology of spinal cord injuries and potential therapeutic approaches. Molecular and Cellular Biochemistry. 2018;**441**(1-2):181-189

[3] Alizadeh A, Dyck SM, Karimi-Abdolrezaee S. Traumatic spinal cord injury: An overview of pathophysiology, models and acute injury mechanisms. Frontiers in Neurology. 2019;**10**:282

[4] Ahuja CS, Wilson JR, Nori S, Kotter MRN, Druschel C, Curt A, et al. Traumatic spinal cord injury. Nature Reviews. Disease Primers. 2017;**3**:17018

[5] Gu Y, Cheng X, Huang X, Yuan Y, Qin S, Tan Z, et al. Conditional ablation of reactive astrocytes to dissect their roles in spinal cord injury and repair. Brain, Behavior, and Immunity. 2019;**80**:394-405

[6] Anderson MA, Burda JE, Ren Y, Ao Y, O'Shea TM, Kawaguchi R, et al. Astrocyte scar formation aids central nervous system axon regeneration. Nature. 2016;**532**(7598):195-200

[7] Okada S, Hara M, Kobayakawa K, Matsumoto Y, Nakashima Y. Astrocyte reactivity and astrogliosis after spinal cord injury. Neuroscience Research. 2018;**126**:39-43

[8] Liberman AC, Trias E, da Silva Chagas L, Trindade P, Dos Santos Pereira M, Refojo D, et al. Neuroimmune and inflammatory signals in complex disorders of the central nervous system. Neuroimmunomodulation. 2018;**25**(5-6):246-270

[9] Nizamutdinov D, Shapiro LA. Overview of traumatic brain injury: An immunological context. Brain Sciences. 2017;**7**(1):11

[10] Matsuda M, Huh Y, Ji RR. Roles of inflammation, neurogenic inflammation, and neuroinflammation in pain. Journal of Anesthesia. 2019;**33**(1):131-139

[11] Chavan SS, Pavlov VA, Tracey KJ. Mechanisms and therapeutic relevance of neuro-immune communication. Immunity. 2017;**46**(6):927-942

[12] Dzyubenko E, Manrique-Castano D, Kleinschnitz C, Faissner A, Hermann DM. Role of immune responses for extracellular matrix remodeling in the ischemic brain. Therapeutic Advances in Neurological Disorders. 2018;**11**:1756286418818092

[13] Pavlov VA, Chavan SS, Tracey KJ. Molecular and functional neuroscience in immunity. Annual Review of Immunology. 2018;**36**:783-812

[14] Shi Y, Wang Y, Li Q, Liu K, Hou J, Shao C. Immunoregulatory mechanisms of mesenchymal stem and stromal cells in inflammatory diseases. Nature Reviews. Nephrology. 2018;**14**(8):493-507

[15] Yang QQ, Zhou JW. Neuroinflammation in the central nervous system: Symphony of glial cells. Glia. 2019;**67**(6):1017-1035

[16] Estrada V, Brazda N, Schmitz C, Heller S, Blazyca H, Martini R, et al. Long-lasting significant functional improvement in chronic severe spinal cord injury following scar resection and polyethylene glycol implantation. Neurobiology of Disease. 2014;**67**:165-179

[17] Raposo C, Schwartz M. Glial scar and immune cell involvement

in tissue remodeling and repair following acute CNS injuries. Glia. 2014;**62**(11):1895-1904

[18] Shechter R, Raposo C, London A, Sagi I, Schwartz M. The glial scar-monocyte interplay: A pivotal resolution phase in spinal cord repair. PLoS One. 2011;**6**(12):e27969

[19] Jassam YN, Izzy S, Whalen M, McGavern DB, El Khoury J. Neuroimmunology of traumatic brain injury: Time for a paradigm shift. Neuron. 2017;**95**(6):1246-1265

[20] Rouanet C, Reges D, Rocha E, Gagliardi V, Silva GS. Traumatic spinal cord injury: Current concepts and treatment update. Arquivos de Neuro-Psiquiatria. 2017;**75**(6):387-393

[21] Oyinbo CA. Secondary injury mechanisms in traumatic spinal cord injury: A nugget of this multiply cascade. Acta Neurobiologiae Experimentalis (Wars). 2011;**71**(2):281-299

[22] Zhou X, He X, Ren Y. Function of microglia and macrophages in secondary damage after spinal cord injury. Neural Regeneration Research. 2014;**9**(20):1787-1795

[23] Wang X, Cao K, Sun X, Chen Y, Duan Z, Sun L, et al. Macrophages in spinal cord injury: Phenotypic and functional change from exposure to myelin debris. Glia. 2015;**63**(4):635-651

[24] Guo L, Rolfe AJ, Wang X, Tai W, Cheng Z, Cao K, et al. Rescuing macrophage normal function in spinal cord injury with embryonic stem cell conditioned media. Molecular Brain. 2016;**9**(1):48

[25] Ren Y, Young W. Managing inflammation after spinal cord injury through manipulation of macrophage function. Neural Plasticity. 2013;**2013**:945034

[26] Beck KD, Nguyen HX, Galvan MD, Salazar DL, Woodruff TM, Anderson AJ. Quantitative analysis of cellular inflammation after traumatic spinal cord injury: Evidence for a multiphasic inflammatory response in the acute to chronic environment. Brain. 2010;**133**(Pt 2):433-447

[27] Bollaerts I, Van Houcke J, Andries L, De Groef L, Moons L. Neuroinflammation as fuel for axonal regeneration in the injured vertebrate central nervous system. Mediators of Inflammation. 2017;**2017**:9478542

[28] Faulkner JR, Herrmann JE, Woo MJ, Tansey KE, Doan NB, Sofroniew MV. Reactive astrocytes protect tissue and preserve function after spinal cord injury. The Journal of Neuroscience. 2004;**24**(9):2143-2155

[29] Goritz C, Dias DO, Tomilin N, Barbacid M, Shupliakov O, Frisen J. A pericyte origin of spinal cord scar tissue. Science. 2011;**333**(6039):238-242

[30] Yuan YM, He C. The glial scar in spinal cord injury and repair. Neuroscience Bulletin. 2013;**29**(4):421-435

[31] Matthews MA, St. Onge MF, Faciane CL, Gelderd JB. Axon sprouting into segments of rat spinal cord adjacent to the site of a previous transection. Neuropathology and Applied Neurobiology. 1979;**5**(3):181-196

[32] Adams KL, Gallo V. The diversity and disparity of the glial scar. Nature Neuroscience. 2018;**21**(1):9-15

[33] Mukhamedshina YO, Akhmetzyanova ER, Martynova EV, Khaiboullina SF, Galieva LR, Rizvanov AA. Systemic and local cytokine profile following spinal cord injury in rats: A multiplex analysis. Frontiers in Neurology. 2017;**8**:581

[34] Liu C, Shi Z, Fan L, Zhang C, Wang K, Wang B. Resveratrol improves

neuron protection and functional recovery in rat model of spinal cord injury. Brain Research. 2011;**1374**:100-109

[35] Rostami A, Ciric B. Role of Th17 cells in the pathogenesis of CNS inflammatory demyelination. Journal of the Neurological Sciences. 2013;**333**(1-2):76-87

[36] Cregg JM, DePaul MA, Filous AR, Lang BT, Tran A, Silver J. Functional regeneration beyond the glial scar. Experimental Neurology. 2014;**253**:197-207

[37] Noble LJ, Donovan F, Igarashi T, Goussev S, Werb Z. Matrix metalloproteinases limit functional recovery after spinal cord injury by modulation of early vascular events. The Journal of Neuroscience. 2002;**22**(17):7526-7535

[38] Hanamsagar R, Hanke ML, Kielian T. Toll-like receptor (TLR) and inflammasome actions in the central nervous system. Trends in Immunology. 2012;**33**(7):333-342

[39] Sticozzi C, Belmonte G, Meini A, Carbotti P, Grasso G, Palmi M. IL-1beta induces GFAP expression in vitro and in vivo and protects neurons from traumatic injury-associated apoptosis in rat brain striatum via NFkappaB/Ca^{2+}-calmodulin/ERK mitogen-activated protein kinase signaling pathway. Neuroscience. 2013;**252**:367-383

[40] Miller SJ. Astrocyte heterogeneity in the adult central nervous system. Frontiers in Cellular Neuroscience. 2018;**12**:401

[41] DiProspero NA, Meiners S, Geller HM. Inflammatory cytokines interact to modulate extracellular matrix and astrocytic support of neurite outgrowth. Experimental Neurology. 1997;**148**(2):628-639

[42] Miljkovic D, Timotijevic G, Mostarica SM. Astrocytes in the tempest of multiple sclerosis. FEBS Letters. 2011;**585**(23):3781-3788

[43] Tang X, Davies JE, Davies SJ. Changes in distribution, cell associations, and protein expression levels of NG2, neurocan, phosphacan, brevican, versican V2, and tenascin-C during acute to chronic maturation of spinal cord scar tissue. Journal of Neuroscience Research. 2003;**71**(3):427-444

[44] Dias DO, Kim H, Holl D, Werne Solnestam B, Lundeberg J, Carlen M, et al. Reducing pericyte-derived scarring promotes recovery after spinal cord injury. Cell. 2018;**173**(1):153-165

[45] Klapka N, Muller HW. Collagen matrix in spinal cord injury. Journal of Neurotrauma. 2006;**23**(3-4):422-435

[46] Wanner IB, Anderson MA, Song B, Levine J, Fernandez A, Gray-Thompson Z, et al. Glial scar borders are formed by newly proliferated, elongated astrocytes that interact to corral inflammatory and fibrotic cells via STAT3-dependent mechanisms after spinal cord injury. The Journal of Neuroscience. 2013;**33**(31):12870-12886

[47] Pekny M, Johansson CB, Eliasson C, Stakeberg J, Wallen A, Perlmann T, et al. Abnormal reaction to central nervous system injury in mice lacking glial fibrillary acidic protein and vimentin. The Journal of Cell Biology. 1999;**145**(3):503-514

[48] Okada S, Nakamura M, Katoh H, Miyao T, Shimazaki T, Ishii K, et al. Conditional ablation of Stat3 or Socs3 discloses a dual role for reactive astrocytes after spinal cord injury. Nature Medicine. 2006;**12**(7):829-834

[49] Nathan FM, Li S. Environmental cues determine the fate of astrocytes after spinal cord injury. Neural Regeneration Research. 2017;**12**(12):1964-1970

[50] Schachtrup C, Ryu JK, Helmrick MJ, Vagena E, Galanakis DK, Degen JL, et al. Fibrinogen triggers astrocyte scar formation by promoting the availability of active TGF-beta after vascular damage. The Journal of Neuroscience. 2010;**30**(17):5843-5854

[51] Rolls A, Shechter R, London A, Segev Y, Jacob-Hirsch J, Amariglio N, et al. Two faces of chondroitin sulfate proteoglycan in spinal cord repair: A role in microglia/macrophage activation. PLoS Medicine. 2008;**5**(8):e171

[52] Horky LL, Galimi F, Gage FH, Horner PJ. Fate of endogenous stem/progenitor cells following spinal cord injury. The Journal of Comparative Neurology. 2006;**498**(4):525-538

[53] Meletis K, Barnabe-Heider F, Carlen M, Evergren E, Tomilin N, Shupliakov O, et al. Spinal cord injury reveals multilineage differentiation of ependymal cells. PLoS Biology. 2008;**6**(7):e182

[54] Sarafian TA, Montes C, Imura T, Qi J, Coppola G, Geschwind DH, et al. Disruption of astrocyte STAT3 signaling decreases mitochondrial function and increases oxidative stress in vitro. PLoS One. 2010;**5**(3):e9532

[55] Smith GM, Strunz C. Growth factor and cytokine regulation of chondroitin sulfate proteoglycans by astrocytes. Glia. 2005;**52**(3):209-218

[56] Xiao Q, Du Y, Wu W, Yip HK. Bone morphogenetic proteins mediate cellular response and, together with Noggin, regulate astrocyte differentiation after spinal cord injury. Experimental Neurology. 2010;**221**(2):353-366

[57] Duchossoy Y, Horvat JC, Stettler O. MMP-related gelatinase activity is strongly induced in scar tissue of injured adult spinal cord and forms pathways for ingrowing neurites. Molecular and Cellular Neurosciences. 2001;**17**(6):945-956

[58] Zhang H, Chang M, Hansen CN, Basso DM, Noble-Haeusslein LJ. Role of matrix metalloproteinases and therapeutic benefits of their inhibition in spinal cord injury. Neurotherapeutics. 2011;**8**(2):206-220

[59] Codeluppi S, Svensson CI, Hefferan MP, Valencia F, Silldorff MD, Oshiro M, et al. The Rheb-mTOR pathway is upregulated in reactive astrocytes of the injured spinal cord. The Journal of Neuroscience. 2009;**29**(4):1093-1104

[60] Liu B, Neufeld AH. Activation of epidermal growth factor receptor causes astrocytes to form cribriform structures. Glia. 2004;**46**(2):153-168

[61] Kohta M, Kohmura E, Yamashita T. Inhibition of TGF-beta1 promotes functional recovery after spinal cord injury. Neuroscience Research. 2009;**65**(4):393-401

[62] Herrmann JE, Imura T, Song B, Qi J, Ao Y, Nguyen TK, et al. STAT3 is a critical regulator of astrogliosis and scar formation after spinal cord injury. The Journal of Neuroscience. 2008;**28**(28):7231-7243

[63] Hamada Y, Ikata T, Katoh S, Katoh K, Niwa M, Tsutsumishita Y, et al. Effects of exogenous transforming growth factor-beta 1 on spinal cord injury in rats. Neuroscience Letters. 1996;**203**(2):97-100

[64] Logan A, Berry M, Gonzalez AM, Frautschy SA, Sporn MB, Baird A. Effects of transforming growth factor beta 1 on scar production in the injured

central nervous system of the rat. The European Journal of Neuroscience. 1994;**6**(3):355-363

[65] Su Z, Yuan Y, Chen J, Zhu Y, Qiu Y, Zhu F, et al. Reactive astrocytes inhibit the survival and differentiation of oligodendrocyte precursor cells by secreted TNF-alpha. Journal of Neurotrauma. 2011;**28**(6):1089-1100

[66] Gris P, Tighe A, Levin D, Sharma R, Brown A. Transcriptional regulation of scar gene expression in primary astrocytes. Glia. 2007;**55**(11):1145-1155

[67] Regis G, Pensa S, Boselli D, Novelli F, Poli V. Ups and downs: The STAT1:STAT3 seesaw of interferon and gp130 receptor signalling. Seminars in Cell & Developmental Biology. 2008;**19**(4):351-359

[68] Shechter R, London A, Varol C, Raposo C, Cusimano M, Yovel G, et al. Infiltrating blood-derived macrophages are vital cells playing an anti-inflammatory role in recovery from spinal cord injury in mice. PLoS Medicine. 2009;**6**(7):e1000113

[69] Sofroniew MV. Astrogliosis. Cold Spring Harbor Perspectives in Biology. 2014;**7**(2):a020420

[70] Karimi-Abdolrezaee S, Billakanti R. Reactive astrogliosis after spinal cord injury—Beneficial and detrimental effects. Molecular Neurobiology. 2012;**46**(2):251-264

[71] Silver J, Miller JH. Regeneration beyond the glial scar. Nature Reviews. Neuroscience. 2004;**5**(2):146-156

[72] Sahni V, Mukhopadhyay A, Tysseling V, Hebert A, Birch D, McGuire TL, et al. BMPR1a and BMPR1b signaling exert opposing effects on gliosis after spinal cord injury. The Journal of Neuroscience. 2010;**30**(5):1839-1855

[73] Pendleton JC, Shamblott MJ, Gary DS, Belegu V, Hurtado A, Malone ML, et al. Chondroitin sulfate proteoglycans inhibit oligodendrocyte myelination through PTPsigma. Experimental Neurology. 2013;**247**:113-121

[74] Trivedi A, Olivas AD, Noble-Haeusslein LJ. Inflammation and spinal cord injury: Infiltrating leukocytes as determinants of injury and repair processes. Clinical Neuroscience Research. 2006;**6**(5):283-292

[75] Rasouli A, Bhatia N, Dinh P, Cahill K, Suryadevara S, Gupta R. Resection of glial scar following spinal cord injury. Journal of Orthopaedic Research. 2009;**27**(7):931-936

[76] Klapka N, Hermanns S, Straten G, Masanneck C, Duis S, Hamers FP, et al. Suppression of fibrous scarring in spinal cord injury of rat promotes long-distance regeneration of corticospinal tract axons, rescue of primary motoneurons in somatosensory cortex and significant functional recovery. The European Journal of Neuroscience. 2005;**22**(12):3047-3058

[77] Schiwy N, Brazda N, Muller HW. Enhanced regenerative axon growth of multiple fibre populations in traumatic spinal cord injury following scar-suppressing treatment. The European Journal of Neuroscience. 2009;**30**(8):1544-1553

[78] Rodriguez-Barrera R, Flores-Romero A, Fernandez-Presas AM, Garcia-Vences E, Silva-Garcia R, Konigsberg M, et al. Immunization with neural derived peptides plus scar removal induces a permissive microenvironment, and improves locomotor recovery after chronic spinal cord injury. BMC Neuroscience. 2017;**18**(1):7

[79] Nakajima H, Uchida K, Guerrero AR, Watanabe S, Sugita D,

Takeura N, et al. Transplantation of mesenchymal stem cells promotes an alternative pathway of macrophage activation and functional recovery after spinal cord injury. Journal of Neurotrauma. 2012;**29**(8):1614-1625

[80] Liu WL, Lee YH, Tsai SY, Hsu CY, Sun YY, Yang LY, et al. Methylprednisolone inhibits the expression of glial fibrillary acidic protein and chondroitin sulfate proteoglycans in reactivated astrocytes. Glia. 2008;**56**(13):1390-1400

[81] Mbori NJ, Chuan XY, Feng QX, Alizada M, Zhan J. Evaluation of the combination of methylprednisolone and Tranilast after spinal cord injury in rat models. Journal of Korean Neurosurgical Association. 2016;**59**(4):334-340

[82] Huang MQ, Cao XY, Chen XY, Liu YF, Zhu SL, Sun ZL, et al. Saikosaponin a increases interleukin-10 expression and inhibits scar formation after sciatic nerve injury. Neural Regeneration Research. 2018;**13**(9):1650-1656

[83] Brambilla R, Bracchi-Ricard V, Hu WH, Frydel B, Bramwell A, Karmally S, et al. Inhibition of astroglial nuclear factor kappaB reduces inflammation and improves functional recovery after spinal cord injury. The Journal of Experimental Medicine. 2005;**202**(1):145-156

[84] Li S, Gu X, Yi S. The regulatory effects of transforming growth factor-beta on nerve regeneration. Cell Transplantation. 2017;**26**(3):381-394

[85] Gaudet AD, Fonken LK. Glial cells shape pathology and repair after spinal cord injury. Neurotherapeutics. 2018;**15**(3):554-577

[86] Nave KA. Myelination and support of axonal integrity by glia. Nature. 2010;**468**(7321):244-252

[87] Assinck P, Duncan GJ, Hilton BJ, Plemel JR, Tetzlaff W. Cell transplantation therapy for spinal cord injury. Nature Neuroscience. 2017;**20**(5):637-647

[88] Oudega M, Xu XM. Schwann cell transplantation for repair of the adult spinal cord. Journal of Neurotrauma. 2006;**23**(3-4):453-467

[89] Kuo HS, Tsai MJ, Huang MC, Chiu CW, Tsai CY, Lee MJ, et al. Acid fibroblast growth factor and peripheral nerve grafts regulate Th2 cytokine expression, macrophage activation, polyamine synthesis, and neurotrophin expression in transected rat spinal cords. The Journal of Neuroscience. 2011;**31**(11):4137-4147

[90] Do-Thi A, Perrin FE, Desclaux M, Saillour P, Amar L, Privat A, et al. Combination of grafted Schwann cells and lentiviral-mediated prevention of glial scar formation improve recovery of spinal cord injured rats. Journal of Chemical Neuroanatomy. 2016;**76** (Pt A):48-60

[91] Lin L, Lin H, Bai S, Zheng L, Zhang X. Bone marrow mesenchymal stem cells (BMSCs) improved functional recovery of spinal cord injury partly by promoting axonal regeneration. Neurochemistry International. 2018;**115**:80-84

[92] Xiong LL, Liu F, Lu BT, Zhao WL, Dong XJ, Liu J, et al. Bone marrow mesenchymal stem-cell transplantation promotes functional improvement associated with CNTF-STAT3 activation after hemi-sectioned spinal cord injury in tree shrews. Frontiers in Cellular Neuroscience. 2017;**11**:172

[93] Okuda A, Horii-Hayashi N, Sasagawa T, Shimizu T, Shigematsu H, Iwata E, et al. Bone marrow stromal cell sheets may promote axonal regeneration and functional recovery with suppression of glial scar formation

after spinal cord transection injury in rats. Journal of Neurosurgery. Spine. 2017;**26**(3):388-395

[94] Jeong SR, Kwon MJ, Lee HG, Joe EH, Lee JH, Kim SS, et al. Hepatocyte growth factor reduces astrocytic scar formation and promotes axonal growth beyond glial scars after spinal cord injury. Experimental Neurology. 2012;**233**(1):312-322

[95] Li Y, Decherchi P, Raisman G. Transplantation of olfactory ensheathing cells into spinal cord lesions restores breathing and climbing. The Journal of Neuroscience. 2003;**23**(3):727-731

[96] Gong S, Jin H, Zhang D, Zou W, Wang C, Li Z, et al. The therapeutic effects after transplantation of whole-layer olfactory mucosa in rats with optic nerve injury. BioMed Research International. 2018;**2018**:6069756

[97] Wright AA, Todorovic M, Tello-Velasquez J, Rayfield AJ, St. John JA, Ekberg JA. Enhancing the therapeutic potential of olfactory ensheathing cells in spinal cord repair using neurotrophins. Cell Transplantation. 2018;**27**(6):867-878

[98] Yao R, Murtaza M, Velasquez JT, Todorovic M, Rayfield A, Ekberg J, et al. Olfactory ensheathing cells for spinal cord injury: Sniffing out the issues. Cell Transplantation. 2018;**27**(6):879-889

[99] Khankan RR, Griffis KG, Haggerty-Skeans JR, Zhong H, Roy RR, Edgerton VR, et al. Olfactory ensheathing cell transplantation after a complete spinal cord transection mediates neuroprotective and Immunomodulatory mechanisms to facilitate regeneration. The Journal of Neuroscience. 2016;**36**(23):6269-6286

[100] Lopez-Vales R, Garcia-Alias G, Fores J, Vela JM, Navarro X, Verdu E. Transplanted olfactory ensheathing cells modulate the inflammatory response in the injured spinal cord. Neuron Glia Biology. 2004;**1**(3):201-209

[101] Tsintou M, Dalamagkas K, Seifalian AM. Advances in regenerative therapies for spinal cord injury: A biomaterials approach. Neural Regeneration Research. 2015;**10**(5):726-742

[102] Deng LX, Hu J, Liu N, Wang X, Smith GM, Wen X, et al. GDNF modifies reactive astrogliosis allowing robust axonal regeneration through Schwann cell-seeded guidance channels after spinal cord injury. Experimental Neurology. 2011;**229**(2):238-250

[103] Ikeda H, Wu GY, Wu CH. Evidence that an iron chelator regulates collagen synthesis by decreasing the stability of procollagen mRNA. Hepatology. 1992;**15**(2):282-287

[104] Kawano H, Li HP, Sango K, Kawamura K, Raisman G. Inhibition of collagen synthesis overrides the age-related failure of regeneration of nigrostriatal dopaminergic axons. Journal of Neuroscience Research. 2005;**80**(2):191-202

[105] Davies JE, Tang X, Denning JW, Archibald SJ, Davies SJ. Decorin suppresses neurocan, brevican, phosphacan and NG2 expression and promotes axon growth across adult rat spinal cord injuries. The European Journal of Neuroscience. 2004;**19**(5):1226-1242

[106] Niclou SP, Ehlert EM, Verhaagen J. Chemorepellent axon guidance molecules in spinal cord injury. Journal of Neurotrauma. 2006;**23**(3-4):409-421

[107] Lu P, Jones LL, Tuszynski MH. Axon regeneration through scars and into sites of chronic spinal cord injury. Experimental Neurology. 2007;**203**(1):8-21

Chapter 7

The Genetic Aspects of Behçet's Disease: Role of Cytokine Genes Polymorphisms

Abdulrahman Al Asmari and Misbahul Arfin

Abstract

Behçet's disease (BD) is a complex, multisystemic inflammatory disorder characterized by recurrent oral aphthous ulcers, ocular symptoms, skin lesions, and genital ulcerations. The etiology of BD is not yet clear though various factors including environmental, genetic and immunological ones have been implicated. Genetic predisposition is a major factor in disease susceptibility and multiple host genetic factors have been suggested to be involved in the development of BD. In addition to the positive association of HLAB*51, recent studies report additional independent associations in the non HLA loci. Single nucleotide polymorphisms (SNPs) in various genes including cytokines have been implicated in susceptibility to BD. However, the results are inconsistent and variation are found in several ethnic populations. Therefore, further genetic studies on BD patients of different ethnicity and genes associated with immunity are expected to elucidate BD pathogenesis and will contribute to the development of more targeted therapies and biomarkers.

Keywords: Behçet's disease, genetics, cytokines, TNF, interleukin, polymorphism

1. Introduction

Behçet's disease (BD; MIM 109650) is a multisystemic inflammatory disorder characterized by recurrent oral aphthous ulcers, ocular symptoms, skin lesions, and genital ulcerations. BD has many features in common with systemic vasculitis. The prevalence of BD varies, it is more prevalent in the Far East, the Mediterranean and the Middle Eastern countries along the ancient Silk Road [1–3]. The highest prevalence has been reported in Turkey (80–420 cases per 100,000) followed by Israel (146.4), China (110), Iran (80), Korea (30.2), Japan (22), Saudi Arabia (20), Iraq (17), Morocco (15), and Egypt (7.5) cases per 100,000 [1, 3, 4].

Clinical and immunological understandings of the disease suggest that BD is a cornerstone between autoimmune and inflammatory disease [5]. Due to effectiveness of immunosuppressives [6] and involvement of human heat-shock protein 60 (HSP60) [7], it is considered as an autoimmune disease. While on the basis of lack of antigen-specific T-cells or significant high-titer auto-antibodies, insignificant involvement of histocompatibility complex (MHC) class I molecules together with unprovoked recurrent inflammation episodes mainly caused by neutrophils [8], the association of the M694V MEFV mutation with its susceptibility and the therapeutic effectiveness of colchicine, BD is classified as an auto-inflammatory disease [4].

IntechOpen

Although it is thought that common environmental factors such as infections or exposures to toxins or to specific immunogens contribute to BD, development of disease is believed to occur only in genetically predisposed hosts. BD is a complex disease and different patients experience different symptoms. The etiology of BD is very complex and it is thought that environmental factors, genetic predisposition and immune dysregulation are involved in the pathogenesis of BD [9–14]. The wide range of disease prevalence observed among different geographic locales is likely a result of differences in both environment and genetics. The aim of this study is to highlight the genetic aspect of BD with emphasis on the role of cytokine genes polymorphisms in the susceptibility/etiopathogenesis of BD.

2. Genetic aspect of Behçet's disease

Genetic predisposition is a major factor in disease susceptibility and multiple host genetic factors have been suggested to be involved in the development of BD. The association of HLAB*51 with BD susceptibility has been confirmed in several populations since it was discovered more than four decades ago however, recent studies indicate association in the major histocompatibility complex class I region and several non HLA loci also. Class I alleles, HLA-A*26, -B*15, -B*27, and -B*57, have been reported as independent risk factors for Behçet's disease while HLA- A*03 and -B*49 are protective for it [15].

The candidate gene approach has been useful in identifying susceptibility and severity genes in BD. Single nucleotide polymorphisms (SNPs) in various genes (IL-10, TNF-α, TNF-β, STAT4, IL23R, CD40, CCR1/CCR3, STAT3, MCP-1, TGFBR3, FCRL3, SUMO4, UBAC2) have been implicated in susceptibility to BD. However, the results are inconsistent and variation are found in several ethnic populations. Genome-wide association studies have also identified associations with IL23R–IL12RB2, IL10, STAT4, CCR1-CCR3, KLRC4, ERAP1, TNFAIP3, and FUT2 loci [15]. Moreover rare mutations in IL23R, TLR4, NOD2, and MEFVr genes have been found to be linked with BD pathogenesis by targeted next-generation sequencing.

The variations in the mRNA expression/gene function indicate the role of the risk alleles in the pathogenesis of disease. Several susceptibility genes, which may regulate the immune reaction, have been found to be associated with BD. However, the precise mechanism of these genes in the development of BD is currently unknown [10, 11, 16]. The genes identified are involved in both innate and adaptive immunity and support the idea that polarization in Th1/Th17 pathway plays a critical role in BD pathogenesis. Commonalities of susceptibility genes with other immune-related diseases/inflammatory disorders shows shared features of immune related diseases with BD. The interaction between genetic factors and environmental factors has also been suggested in several recent studies.

Cytokines are believed to mediate inflammation in BD [17, 18]. Various studies have found increased levels of tumor necrosis factor (TNF)-α and decreased levels of interleukin (IL)-10 in the serum and active lesions of BD patients and suggested that these cytokines play a significant role in the immune response, pathogenesis and activity in BD [12, 19–24].

Cytokines play critical roles in the pathogenesis of BD, since they mediate many of the effector and regulatory functions of immune and inflammatory responses [14, 17]. Genetic polymorphisms in several cytokine genes have been described and demonstrated to influence gene transcription, leading to inter individual variations in cytokine production. It has been suggested that genetic polymorphisms that regulate the production of certain cytokines are important determinants of susceptibility to BD and its some of the clinical and laboratory features [14, 25, 26]. BD has been considered to be a typical Th1-mediated inflammatory disease, characterized by elevated levels of Th1 cytokines such as IFN-γ, IL-2, and TNF-α. Recently it has been reported that Th1- and Th17-related cytokines and signaling molecules participated in BD pathogenesis [27, 28].

A number of studies reported that the levels of T helper Type 1 cytokines are increased in sera of the patients with BD. Some studies have shown that the maximal capacity of cytokine production varies among individuals and correlate with single nucleotide polymorphism in various cytokine genes [29–32]. However the results of the association of cytokines genes polymorphism with susceptibility and pathogenesis of BD are inconsistent and further studies involving different ethnic populations have been suggested [14].

3. Tumor necrosis factor (TNF)-α polymorphisms

Besides HLA-B51 molecules, SNPs in TNF genes have been implicated in susceptibility to BD [14, 33–37]. TNF-α is a pro-inflammatory cytokine and involved in regulation of the immune response. It is encoded in the Class III region of the HLA complex adjacent to HLA-B. TNF-α mediates the activation of macrophages and apoptosis and it is involved in recurrent inflammatory episodes in BD patients [23, 38]. Many studies have suggested it as both positional and functional candidate gene in the onset and progression of BD [14, 33–35, 39].

Promoter polymorphism of TNF-α (−308G/A) and intronic polymorphism TNF-β (252A/G) have been associated with variations in the level of circulating TNF-α [40]. TNF-α (−308G/A) polymorphism (rs1800629) results into a less common allele-A (allele 2) which leads to increased TNF-α production in vitro [41] and higher rate of TNF-α transcription than wild type allele-G (allele 1). Allele-A produces 6–7 fold higher levels of TNF-α transcription [42–44]. TNF-α production and expression is regulated by single nucleotide polymorphisms (SNPs) in TNF-α gene [25, 42]. Several SNPs in TNF-α gene have been associated BD in different ethnic groups [14, 36, 45–47]. The outcome of various studies on association between BD and SNPs of TNF-α in different ethnic groups are summarized in **Tables 1–6**.

3.1 TNF-α (−308 G/A) polymorphism

A number of studies has determined the relationship between the -308A/G polymorphism and BD with inconsistent results (**Table 1**). The genotype GA and allele-A are associated with susceptibility of BD in Saudis [14] while genotype GG and allele G are associated with its susceptibility in Korean patients [33]. On the other hand no association is found in Caucasoid [39], Iranian [48], Iranian (Azeri Turkish) [54], Korean [49, 50] Lebanese [51], Tunisian [46], Turkish [38, 45, 52, 53] and Moroccan [47].

Two independent meta-analysis have revealed an association between −308A and BD risk in the overall [34, 35] however, stratification by ethnicity indicates that the −308A allele is significantly associated with BD risk in the Asian population [35].

3.2 TNF-α (−238 A/G) polymorphism

The TNF-238A/G polymorphism has been studied in BD patients from different ethical populations and several reports are available on the association between the TNF−238A/G polymorphism and BD risk with contrast results. Genotype AA is found to be associated with BD in Turkish population [55] whereas genotype GG is associated with BD in Iranian patients [48]. Other individual studies on German

Population	Case/controls	Genotype/allele/polymorphism	Association	Reference
Saudi	61/211	GA/A	Susceptible	[14]
Korean	254/344	GG/G	Susceptible	[33]
Korean	94/94	GA polymorphism	No association	[49]
Korean	115/114	GA polymorphism	No association	[50]
Turkish	99/96	GA polymorphism	No association	[52]
Turkish	107/102	GA polymorphism	No association	[38]
Turkish	97/127	GA polymorphism	No association	[53]
Turkish	102/102	GA polymorphism	No association	[45]
Lebanese	48/90	GA polymorphism	No association	[51]
Iranian	147/137	GA polymorphism	No association	[48]
Iranian (Azeri Turkish)	53/79	GA polymorphism	No association	[54]
Moroccan	120/112	GA polymorphism	No association	[47]
Tunisian	89/157	GA polymorphism	No association	[46]
Caucasoid	133/354	GA polymorphism	No association	[39]
Meta-analysis	–	GA polymorphism	No association	[36]
Meta-analysis	1372/1754	GA polymorphism	Susceptible	[34]
Asian*	1232/1397	A-allele	Susceptible	[35]

**Meta-analysis.*

Table 1.
Association of TNF-α-308 polymorphism with BD susceptibility.

[56], Iranian (Azeri Turkish) [57], Korean [33, 50], Lebanese [51], Moroccan [47] and Turkish [38, 56] BD patients show no association of TNF−238A/G polymorphism with susceptibility of BD (**Table 2**). However two meta-analysis indicates that allele-A is associated with BD susceptibility [35, 36]. In the subgroup analysis by ethnicity, Zhang et al. [35] suggests that the BD cases has a significant higher frequency of A in the Caucasian than that in the controls.

3.3 TNF-α (−1031 C/T) polymorphism

There is a strong evidence indicating the role of TNF-α (−1031 C/T) Polymorphism with BD susceptibility. A number of studies examined the association of TNF-α −1031C/T with BD in different populations. Results indicate, a significant association between the TNF-α −1031C/T polymorphism and BD susceptibility in

Population	Case/controls	Genotype/allele/polymorphism	Association	Reference
Turkish	107/102	A/G polymorphism	No association	[38]
Turkish	80/105	AA	Susceptible	[55]
Turkish	30/20	A/G polymorphism	No association	[56]
Korean	254/344	A/G polymorphism	No association	[33]
Korean	115/114	A/G polymorphism	No association	[50]
German	92/51	A/G polymorphism	No association	[56]
Lebanese	48/90	A/G polymorphism	No association	[51]
Moroccan	120/112	A/G polymorphism	No association	[47]
Iranian	150/140	GG	Susceptible	[48]
Iranian (Azeri Turkish)	64/101	A/G polymorphism	No association	[57]
Meta-analysis	–	A-allele	Susceptible	[36]
Caucasian*	842/938	A-allele	Susceptible	[35]

**Meta-analysis.*

Table 2.
Association of TNF-α −238 polymorphism with BD susceptibility.

Turkish [29, 58], Iranian (Azeri Turkish) [54], Korean [33], and Tunisian patients [46] (**Table 3**). Touma et al. [36] in a meta-analysis, identified a significant associations between the −1031C/T polymorphisms and BD risk. Stratifying by ethnicity, in another meta-analysis a significant association in the Caucasian population is noticed [35]. Radouane et al. [47] suggested that TNF-1031C constitutes a susceptibility allele for BD in Moroccan, especially with genital ulcers.

In contrast, Chang et al. [50] discovered no significant difference in the allele frequency of TNF-α −1031C/T between patients with BD and controls in a Korean population. There was no significant association of this polymorphism in Lebanese BD patients also [51]. Moreover the analysis of the influences this polymorphism on various clinical manifestations of BD showed that TNF-α −1031C is not related to the presence of clinical features, such as oral and genital ulceration and uveitis.

3.4 TNF-α (−857 T/C) polymorphism

A number of studies has been focused on the association between the TNF-α-857T/C polymorphism and BD risk (**Table 4**). Two independent studies indicate an association of TNF-α-857T/C polymorphism with BD susceptibility in Korean and Iranian (Azeri Turkish) Cohort [33, 57]. Other studies performed on Korean, Lebanese, and Moroccan BD patients show no significant association of this polymorphism and BD susceptibility [47, 50, 51]. However a meta-analysis suggested that T-allele of TNF-α-857T/C polymorphism is associated with BD susceptibility [36]. Later on another meta-analysis also indicated that this association is a significant risk factor in Asian population [35].

Population	Case/controls	Genotype/allele/polymorphism	Association	Reference
Turkish	99/103	C/T polymorphism	Susceptible	[29]
Turkish	82/77	CC	Susceptible	[58]
Iranian (Azeri Turkish)	53/79	C-allele	Susceptible	[54]
Korean	254/344	C/T polymorphism	Susceptible	[33]
Korean	115/114	C/T polymorphism	No association	[50]
Tunisian	89/157	C-allele	Susceptible	[46]
Lebanese	48/90	C/T polymorphism	No association	[51]
Caucasoid (UK)	133/354	C/T polymorphism	Susceptible	[39]
Moroccan	120/112	C-allele	Susceptible	[47]
Meta-analysis		C- allele	Susceptible	[36]
Caucasian*	738/964	C- allele	Susceptible	[35]

**Meta-analysis.*

Table 3.
Association of TNF-α –1031 polymorphism with BD susceptibility.

Population	Case/controls	Genotype/allele/polymorphism	Association	Reference
Korean	254/344	C-allele	Susceptible	[33]
Korean	115/114	T/C polymorphism	No association	[50]
Lebanese	48/90	T/C polymorphism	No association	[51]
Moroccan	120/112	T/C polymorphism	No association	[47]
Iranian (Azeri Turkish)	64/101	C-allele	Susceptible	[57]
Meta-analysis		T-allele	Susceptible	[36]
Asian*	533/660	T/-allele	Susceptible	[35]

**Meta-analysis.*

Table 4.
Association of TNF-α –857 polymorphism and BD in various populations.

3.5 TNF-α (−863 A/C) polymorphism

TNF-863A/C polymorphisms has been studied in Korean, Moroccan and Lebanese BD patients. Results of these studies indicated that there is no significant association of polymorphism with BD susceptibility (**Table 5**) [47, 50, 51] though one study suggested an association in Korean patients [33]. However two independent meta-analysis also did not find any significant role of this polymorphism in BD susceptibility [35, 36].

3.6 TNF-α (−376 A/G) polymorphism

Three reports are available on −376A/G polymorphism in BD, two studies were performed in Turkish while the third one in Moroccan patients. These studies identified no significant association with BD risk (**Table 6**). The results of the meta-analysis

Population	Case/controls	Genotype/allele/polymorphism	Association	Reference
Korean	254/344	A/C polymorphism	Susceptible	[33]
Korean	115/114	A/C polymorphism	No association	[50]
Moroccan	120/112	A/C polymorphism	No association	[47]
Lebanese	48/90	A/C polymorphism	No association	[51]
Meta-analysis	–	A/C polymorphism	No association	[36]
Asian*	486/560	A/C polymorphism	No association	[35]

**Meta-analysis.*

Table 5.
Association of TNF-α–863 polymorphism with BD in various populations.

Population	Case/controls	Genotype/allele/polymorphism	Association	Reference
Turkish	99/96	A/G polymorphism	No association	[52]
Turkish	107/102	A/G polymorphism	No association	[38]
Moroccan	120/112	A/G polymorphism	No association	[47]

Table 6.
Association of TNF-α–376 polymorphism with BD susceptibility.

by Zhang et al. [35] also showed that the TNF-376A/G polymorphism is not associated with BD susceptibility and this polymorphism does not appear to have a significant association with overall BD risk.

4. TNF-β (–252A/G) polymorphism

TNF-β has been reported to contribute to the susceptibility of some inflammatory and autoimmune diseases [58–62]. Gamma delta T cells of BD patients produce higher levels of TNF-β than those of healthy controls [63, 64]. A polymorphism in the intron 1 of TNF-β has been associated with higher TNF-α and TNF-β production. TNF-β (+252A/G) polymorphism (rs909253) contains a Guanine (G) on one allele and an adenine (A) on the alternate allele. TNF-β +252G allele is defined as mutant allele and known as TNF-β*1 (allele-1). This mutant allele-1 is associated with increased levels of TNF-α and TNF-β [65, 66].

To the best of our knowledge five studies have been focused on TNF-β (+252A/G) polymorphism and BD (**Table 7**). The results of three studies in Saudi, Korean and Tunisian BD patients indicated that TNF-β (+252A/G) polymorphism has no significant association with BD susceptibility. However one report from

Population	Case/controls	Genotype/allele/polymorphism	Association	Reference
Saudi	61/211	AG polymorphism	No association	[14]
Tunisian	89/157	AG polymorphism	No association	[46]
Korean	94/94	AG polymorphism	No association	[49]
Middle eastern	102/115	A-allele	Susceptible	[67]
Japanese	79/75	A-allele	Susceptible	[68]

Table 7.
Association of TNF-β–252 polymorphism with BD susceptibility.

Palestinian and Jordanian populations indicates that the frequency TNF-β +252 A allele (allele-2) is increased in BD cases compared to controls [67]. On the basis of strong linkage disequilibrium found between HLA-B*51 and allele-A of TNF-β (+252A/G) polymorphism it has also been suggested that that both the alleles contribute to BD risk and their co-expression may cause severe eye pathogenicity leading to blindness [67]. Another report by Mizuki et al. [68] shows that the frequency of homozygous genotype (GG) of TNF-β (+252A/G) is significantly decreased in Japanese ocular BD patients than controls.

5. Interleukin (IL) gene polymorphisms

Interleukins are cytokines that mediate communication between cells. Interleukins regulate cell growth, differentiation, and motility. They are particularly important in stimulating immune response, such as inflammation. ILs (IL-1 to IL-38) function and play significant role in various diseases and their expression/production is influenced by the polymorphisms and mutations in their encoding genes [69].

5.1 IL-10 gene polymorphism

IL10 gene encodes IL-10 cytokine which suppresses the production of pro-inflammatory cytokines such as IL-1, IL-6, IL-12, TNF, and interferon gamma (IFN-γ), and inhibits the costimulatory activity of macrophages for T cell and NK cell activation [70]. IL-10 production may be regulated at the transcriptional level and several single nucleotide polymorphisms (SNPs) at the promoter region of IL-10 gene have been shown to be associated with changes in the expression levels of IL-10 [25, 42].

Numerous recent studies have demonstrated an association between BD and SNPs of IL10. Three polymorphisms -1082 A/G (rs1800896), -819 T/C (rs1800871) and -592 A/C (rs1800872) in the promoter region of the IL-10 gene are correlated to the expression level of IL-10. There are inconsistent reports on the association of IL-10-1082 A/G, -819 T/C and -592 A/C polymorphisms and BD (**Tables 8–10**). Two recent studies suggested significant association of genotype GG of IL-10-1082 A/G with BD susceptibility in Saudis [14] and Egyptian [71]. Earlier Wallace et al. [72] showed weak association of genotype AA with BD in UK and middle-eastern cohort. Moreover a meta-analysis also shows that there is a significant association of IL-10-1082 A/G polymorphism with BD susceptibility [34]. While there is no significant association of this polymorphism in Turkish and Iranian BD patients (**Table 8**) [28, 45, 53].

Population	Case/controls	Genotype/allele/polymorphism	Association	Reference
Saudi	61/200	GG	Susceptible	[14]
Egyptian	87/97	GG	Susceptible	[71]
UK+ME	178/295	AA	Weekly associated	[72]
Turkish	97/127	GA polymorphism	No association	[53]
Turkish	102/102	GA polymorphism	No association	[45]
Iranian	150/140	GA polymorphism	No association	[28]
Meta-analysis	199/229	GG+GA	Susceptible	[34]

Table 8.
Association of IL-10-1082 polymorphism with BD susceptibility.

Population	Case/controls	Genotype/allele/polymorphism	Association	Reference
Saudi	61/200	TT	Susceptible	[14]
Chinese	407/679	CT polymorphism	Susceptible	[74]
Chinese Han	718/1753	T-allele	Susceptible	[75]
Algerian	51/96	T-allele	Susceptible	[73]
British	178/295	T-allele	Susceptible	[72]
Turkish	102/102	CT polymorphism	No association	[45]
Turkish	97/127	CT polymorphism	No association	[53]
Turkish	1215/1279	CT polymorphism	No association	[11]
Japanese	611/737	CT polymorphism	No association	[11]
Korean	119/140	CT polymorphism	No association	[11]
Iranian	150/140	CT polymorphism	No association	[28]
Overall mixed	1945/2156	CT polymorphism	Susceptible	[11]
Meta-analysis	2472/2820	CT polymorphism	Susceptible	[34]

Table 9.
Association of IL-10-819 polymorphism with BD susceptibility.

Population	Case/controls	Genotype/allele/polymorphism	Association	Reference
Saudi	61/200	AA	Susceptible	[14]
Algerian	51/96	A-allele	Susceptible	[73]
Iranian	150/140	CA polymorphism	No association	[28]
Iranian (Azeri Turkish)	47/58	A-Allele	Susceptible	[76]
Chinese Han	718/1753	A-Allele	Susceptible	[75]
Spanish	304/313	A-Allele	Susceptible	[77]
Turkish	102/102	CA polymorphism	No association	[45]
Turkish	97/127	CA polymorphism	No association	[53]
Turkish	1215/1279	CA polymorphism	No association	[11]
Japanese	611/737	CA polymorphism	No association	[11]
Korean	119/140	CA polymorphism	No association	[11]
Overall mixed	1945/2156	CA polymorphism	Susceptible	[11]
Meta-analysis	2294/2525	CA polymorphism	Susceptible	[34]

Table 10.
Association of IL-10-592 polymorphism with BD susceptibility.

IL-10-819 T/C polymorphism has been studied in different populations (**Table 9**). Reports indicate that IL-10-819 T/C polymorphism is associated with susceptibility of BD in Algerians [73], British [72], Chinese [74, 75] and Saudi patients [14]. While it is not significantly associated with BD in Turkish [11, 45, 53], Iranian [28], Japanese and Korean patients [11]. However two independent meta-analysis showed that IL-10-819 T/C polymorphism is associated with BD [11, 34]. In a meta-analysis containing of 2472 cases and 2820 controls, Liang et al. [34] suggested that IL-10-819 T/C polymorphism is associated with BD susceptibility.

Available literature shows that 11 studies focused on relationship of IL-10-592 A/C polymorphism and BD risk (**Table 10**). A significant association of this polymorphism has been reported in five studies from different ethnicity namely Algerian [73], Iranian (Azeri Turkish) [76], Chinese Han [75], Saudi [14] and Spanish BD patients [77]. In our study with Saudi patients we found that −592 AA genotypes of IL-10 is significantly associated with susceptibility risk of BD in Saudi patients [14].

On the other hand three studies on Turkish [11, 45, 53], one each on Iranian [28], Japanese and Korean BD patients [11] show no significant association of this polymorphism with BD susceptibility. A meta-analysis containing 2294 patients and 2525 controls suggested that IL-10-592 A/C polymorphism is associated with BD susceptibility [34].

Two independent GWA studies of Turkish and Japanese populations show that IL-10 is among the first two BD susceptibility loci outside the MHC with genome-wide significance [11, 16]. Intronic polymorphism (rs1518111) is associated with BD susceptibility in the Turkish population [16] while promoter polymorphisms in IL-10 gene (rs1800871 and rs1800872) are associated with BD in Japanese [11]. The variant rs1518111 has been replicated in BD patients of British, Greek, Korean and Middle Eastern ethnicity and rs1800872 replicated in Turkish and Korean samples [11, 16]. Recently Wu et al. [74] reported the replication of rs1518111 and rs1800871 variants in BD patients from Han Chinese population. The SNP rs1518111 has also been replicated in in the Iranian population showing association with BD [78]. The data from HapMap Project indicate that these three polymorphisms are in strong linkage disequilibrium in populations from both European and Asian ancestries. The decrease in risk allele A of rs1518111 is associated with decreased IL10 expression in monocytes by 35% compared with the non-risk allele G in Turkish patients with BD.

The homozygous genotype AA of rs1518111 is associated with decrease in IL-10 protein in monocytes and found to be stimulated with Toll-like-receptor ligands, such as lipopolysaccharide or the lipoprotein Pam3Cys and muramyl dipeptide [16]. Talat et al. [71] reported that IL-10 serum levels are lower in BD patients than in controls. Baris et al. [79] suggested that IL-10 polymorphisms can be statistically associated with the disease symptoms and used as prognostic factors.

5.2 IL-1 gene polymorphism

Several cytokine genes may play crucial roles in host susceptibility to Behçet's disease (BD), since the cytokine production capacity varies among individuals and depends on the cytokine gene polymorphisms. Interleukin-1 (IL-1) and the IL-1 receptor (IL-1R) family plays an important role in the pathogenesis of inflammatory diseases. The association of the IL-1 cluster gene polymorphisms with the development of BD has been investigated in several studies.

5.2.1 IL-1α −889C/T polymorphism

Six reports are available on IL-1α -889C/T polymorphism and Behçet's disease (**Table 11**). Out these five studies were performed on Turkish patients [55, 58, 80–82] while one on Iranian BD patients [48]. There was no association of this gene polymorphism and the susceptibility of BD except one study which showed CC genotype to be associated with BD susceptibility in Turkish patients [55].

5.2.2 IL-1β −511C/T polymorphism

Four studies are found which focused to assess the importance of IL-1β −511C/T polymorphism for BD susceptibility (**Table 11**). The comparisons of allele and

Population	Case/controls	Polymorphism	Association	Reference
Turkish	132/106	IL-1α-889C/T	No association	[80]
Turkish	72/163	IL-1α-889C/T	No association	[81]
Turkish	80/105	IL-1α-889C/T	CC associated	[55]
Turkish	57/57	IL-1α-889C/T	No association	[58]
Turkish	97/77	IL-1α-889C/T	No association	[82]
Iranian	150/140	IL-1α-889C/T	No association	[48]
Turkish	132/106	IL-1β-511C/T	No association	[80]
Turkish	80/105	IL-1β-511C/T	CC associated	[55]
Turkish	57/57	IL-1β-511C/T	No association	[58]
Turkish	97/77	IL-1β-511C/T	No association	[82]
Turkish	57/57	IL-1β-3962C/T	No association	[58]
Turkish	80/105	IL-1β-3962C/T	CC associated	[55]
Turkish	97/77	IL-1β-3962C/T	No association	[82]
Iranian	150/140	IL-1β-3962C/T	No association	[48]

Table 11.
Association of IL-1 polymorphisms with BD susceptibility.

genotype failed to detect any statistical association under the random effect model in three studies [58, 80, 82] however one study reported that CC genotype of IL-1β −511C/T polymorphism is associated with BD susceptibility in Turkish patients [55].

5.2.3 IL-1β −3962 C/A polymorphism

Some workers have studied IL-1 β −3962 C/A polymorphism and assessed the effect of the IL-1β −3962C/A polymorphism in the occurrence of BD in Turkish and Iranian patients (**Table 11**). They did not find any significant association between IL-1β −3962 C/A polymorphism with BD between allele and genotype frequencies in Turkish and Iranian BD patients [48, 58, 82] however one study indicates association of IL-1 β −3962 C/A polymorphism with BD in Turkish patients [55].

Ozçimen et al. [82] studied IL-1 cluster gene polymorphisms in Turkish patients with Behçet's disease and suggested that polymorphisms in IL-1β gene may affect host susceptibility to BD. The IL-1β production in the active period has been found to be greater than in the remission period of BD. IL-1β production is considered to be related to posterior segment type attacks of Behçet's disease [83].

Baris et al. [79] observed no significant differences between the groups with respect to the IL-1Ra, IL-1β, IL-2, IL-6 and the IL-10 gene polymorphism distributions and suggested that the IL-1RN2 gene polymorphism is correlated with the presence of articular involvement and the IL-1β gene polymorphism with the presence of an ocular lesion. On the basis of the correlations between the articular involvement and IL-1RN, the ocular involvement and the IL-1β, gene polymorphisms, it has been suggested that these polymorphisms could be statistically associated with the disease symptoms and may be used as prognostic factors [79].

6. Interleukin (IL)-6

IL-6 pleitropic cytokine is involved in immune and inflammatory responses. Various polymorphisms in IL-6 gene have been associated with chronic inflammatory

Population	Case/controls	Genotype/allele/polymorphism	Association	Reference
Tunisian	43/43	G/C polymorphism	No association	[20]
Egyptian	87/97	G/C polymorphism	No association	[71]
Iranian	150/140	GG	Protective	[48]
Turkish/German	121/70	G/C polymorphism	No association	[56]
Turkish	97/127	G/C polymorphism	No association	[53]
Korean	89/123	G/C polymorphism	No association	[90]
Meta-analysis	2065/1159	G/C polymorphism	Decrease the risk	[91]

Table 12.
Association of IL-6 (174G/C) polymorphism with BD susceptibility.

and autoimmune disorders [84–88]. Higher levels of IL-6 and increased expression of IL-6 mRNA have been reported in subjects with active BD [20, 89, 90]. A few studies have focused on the polymorphism of IL-6 −174 G/C in BD patients (**Table 12**). The polymorphism of IL6 −174 G/C does not modulate clinical expression of BD. The single nucleotide polymorphism of the IL-6 does not appear to be associated with BD susceptibility in Egyptian [71], Korean [90], Tunisian [20], Turkish and German patients [53, 56]. The GG genotype of IL-6 −174 G/C polymorphism is protective in Iranian population [48]. It is believed that the scarcity of studies of polymorphism of IL-6 in BD is related to the fact that IL-6 is a pro-inflammatory cytokine of Th2, whereas BD is a Th1 disease. Recently a meta-analysis suggested that IL-6-174 G/C polymorphism decreases the risk of BD [91].

7. IL23R-IL12RB2 polymorphisms

The IL23R–IL12RB2 polymorphisms in Behçet's disease have been subject of several studies in various ethnic populations (**Table 13**). The IL23R–IL12RB2 locus is one of the few loci with genome-wide significance. The SNP rs1495965, located in the intergenic region between IL23R and IL12RB2 is associated with BD in Japanese [11]. Another polymorphism (rs924080) in the intergenic region between IL23R and IL12RB2 has been associated with BD in Turkish patients [16]. However, these association are not replicated in Korean, Middle Eastern Arab, Greek, and British subjects possibly due to small sample size [16].

The rs924080 has been replicated in the Iranian population and major allele is associated with BD [78]. Other polymorphisms rs7539328, rs12119179, rs1495965 have also been associated with BD susceptibility in Iranian patients [92]. Moreover minor allele of IL23R polymorphisms, Arg381Gln in the Turkish population and Gly149Arg in the Japanese population are associated with protection from BD as these variants reduce its ability to respond to IL-23 stimulation [99]. Disease–associated, intergenic non-coding variants (major alleles) are associated with increased expression of IL23R compared with the disease-protective minor alleles [99].

The IL-23 receptor is expressed on the surface of Th17 cells and macrophages. It is encoded by *IL23R* gene. IL-23 is composed of p19 and p40 subunits which is shared with IL-12. IL-23, being a proinflammatory cytokine promotes Th17 cell

Population	Case/controls	Polymorphism	Association	Reference
Iranian	973/637	rs10489629 rs1343151 rs1495965	Susceptible	[78]
Iranian	552/417	rs7539328, rs12119179, rs1495965,	Susceptible	[92]
Korean	369/2000	rs1495965 rs1495966 rs4655535	Susceptible	[93]
Chinese Han	407/421	rs924080 rs11209032	Susceptible	[94]
Chinese Han	1206/2475	rs3024490 rs12141431	Susceptible	[95]
Chinese Han	806/1600	rs3212227	Susceptible	[96]
Chinese Han	27/32	rs17375018	Susceptible	[97]
Algerian	51/96	rs12119179 rs11209032 rs924080	decrease risk	[73]
Egyptian		rs17375018	Susceptible	[13]
Turkish	123/168	rs17375018	Susceptible	[98]
Japanese	612/740	rs1495965	Susceptible	[11]
Turkish	2430/2660	rs924080	Susceptible	[16]

Table 13.
Association of IL23R-IL12RB2 polymorphisms with BD susceptibility.

development and induces the production of IL-1, IL-6, IL-17 and TNF [100]. Th17 cells by producing IL-17 play a significant role in inflammation and autoimmune diseases. Steinman [101] suggested that the disease-associated alleles increase IL-23 receptor expression or signaling compared with the disease-protective alleles. Thus it is evident that the disease-associated variants increase the BD susceptibility by influencing IL23R, however an alternative or additional role to influence expression of the other nearby gene, like IL12RB2, cannot be excluded. IL-12 receptor beta2, a subunit of IL-12 receptor is encoded by *IL12RB2* gene. *IL12RB2* is responsible for high-affinity IL-12 binding and IL-12 dependent signaling, and plays an important role in Th1 cell differentiation. IL-12 has been suggested to be involved in Th1 responses, T cell and NK cell cytotoxicity, and IFN-γ production by T cells and NK cells [102]. As there is no data available on quantitative trait loci for these non-coding variants influencing IL12RB2 or IL23R expression, there is a possibility that their effects are expressed only in a specific cell type or under certain conditions as suggested by Takeuchi et al [15].

Yu et al. [95] performed genome wide association study on 1206 patients with BD and 2475 healthy controls and confirmed the association of IL-10 819 C/T and IL23R IL12RB2/rs924080 with BD. They (loc. cit.) also identified two susceptibility single nucleotide polymorphisms in IL10 and IL23R-IL12RB2 (rs3024490 and rs12141431) with BD in Han Chinese.

8. IL12A polymorphisms

IL12A encodes IL-12p35, a subunit of the heterodimer of IL-12, is known to play a critical role in polarization of the Th1 pathway through differentiation from naïve CD4+ T cells. A variant rs1780546, located in the intergenic region near IL12A has been associated with BD in a Turkish cohort however the association did not achieve genome-wide significance as it is not polymorphic in the Japanese cohort [99]. Recently, Kappen et al. [103] reported an association of rs1780546 with BD susceptibility and showed genome-wide association after meta-analysis with previous Turkish GWAS data. This GWAS was based on 336 cases and 5843 controls in cohorts of mixed ethnicity using linear mixed models to correct for ancestry differences and family structure and/or cryptic relationships. However no report is available on functional aspect of rs1780546 [102].

9. IL-33 gene polymorphism

Members of the IL-1 family play a pivotal role in the inflammatory responses [104]. IL-33 belongs to IL-1 superfamily. Some studies have implicated the IL-33 ligand for the ST2 receptor in the pathogenesis of BD [105–107]. IL-33 is encoded by *IL-33* gene and expressed in epithelial, endothelial, inflammatory and central nervous system cells. IL-33 can function both as a cytokine and as a nuclear factor regulating gene transcription due to the fact that its expression increases in pro-inflammatory conditions. Cells of the affected area play a significant role in the pathogenesis of BD by recruiting, activating and promoting survival of inflammatory cells. Therefore the use of immunosuppressant like azathioprine, cyclosporine, corticosteroids or anti-TNF-α monoclonal antibodies (mAb) which interferes the cytokine network is the basis of BD treatment strategies [6, 20, 108].

The TT variants of rs7044343 and rs11792633 polymorphisms in IL-33 gene are very rare, and the T allele frequencies of these polymorphisms has been reported to be lower in the BD group compared to the controls. The rs7044343 and rs11792633 variants of IL-33 gene are associated with the decreased risk of BD in Turkish cohort. It has been suggested that IL-33 acts a protective role on the pathogenesis of BD [109].

10. Interferon-γ (IFN-γ)

IFN-γ is antiviral, antitumor and immunomodulatory cytokine. It has a critical role in modulating the IL-4, IL-10 and IL-12 cytokine network pathway. It is also considered as a pro-inflammatory cytokine because of its effects on TNF activity. It has been reported that the frequencies of IFN-γ +874A allele and A/A genotype are higher in BD patients than in healthy controls, and individuals with this genotype are more susceptible to the disease [55]. However later studies on Turkish and Iranian patients failed to find any significant association of IFN-γ +874 A/T polymorphism with BD susceptibility (**Table 14**) [28, 53].

Population	Case/controls	Genotype/allele/polymorphism	Association	Reference
Turkish	80/105	AA/A	Susceptible	[55]
Turkish	97/127	A/T polymorphism	No association	[53]
Iranian	150/140	A/T polymorphism	No association	[28]

Table 14.
Association of IFN-γ (+874A/T) polymorphism with BD susceptibility.

11. Transforming growth factor-β1 gene polymorphism

Transforming growth factor-β1 (TGF-β1) is an effective immunosuppressive cytokine. It is produced in response to tissue injury by activated macrophages. TGF-β1 is responsible for inhibition of macrophage activation and modulation of T cell function [110, 111]. It is also involved in tissue fibrosis by increasing the synthesis of extracellular matrix components [112]. The frequency of TGF-β1 codon 10–25 T/C-G/C genotype in Turkish BD patients has been reported to be higher than those of healthy controls [53] while GG genotype has been reported to be susceptible to BD in Iranian cohort (**Table 15**) [28].

Recent studies have focused on the functional relevance of the various genes associated with susceptibility of BD and possible interaction between the genes located within and outside the MHC region [14, 24, 113, 114]. The functional relevance of allele A and genotype GA of TNF-α (308G/A) and association with BD has been indicated in various studies [24, 114]. The increased frequency of allele-A in BD patients is linked with higher levels of TNF-α reported in active BD patients as compared to controls [24, 114].

The pro-inflammatory cytokines induce inflammation and the severity of the inflammatory responses is influenced by the levels of cytokines. The activated macrophages produce higher levels of cytokines affecting not only the severity of the local inflammatory responses but also exert systemic effects. The over-expression of these cytokines is considered to be responsible for the pathogenesis of recurrent BD [33]. TNF-α, a pro-inflammatory cytokine has been suggested to be responsible for the pathogenesis of BD by activating T-cells and neutrophils [115].

On the other hand increased frequency of low producer 1082GG genotype of IL-10 (an anti-inflammatory cytokine) in BD patients may not suppress the TNF-α- activity and resulting inflammatory responses, as IL-10 is known to limit the secretion of pro-inflammatory cytokines, such as TNF-*α* and IL-12 [70]. Moreover the deficiency of IL-10 and resulting prolonged activation of mononuclear cells may lead to an augmented efflux of inflammatory cytokines and further aggravate the severity of BD as IL-10 is a multi-functional cytokine with role in diverse areas of the human immune system [116].

The information regarding the association of various gene polymorphisms will have prognostic value for future clinical observations. Especially the data of TNF-α (−308) polymorphism will provide guideline in anti-TNF-α therapy as patients with GG genotype are better responders to anti-TNF-α treatment than those with AA or GA [117, 118]. However, such genetic associations with BD susceptibility need further validation and investigation in more patients with BD from various ethnic populations, as they may have implications for the development of novel therapies as suggested by Xavier *et al* [78].

Population	Case/controls	Genotype/allele	Association	Reference
Turkish	97/127	T/C-G/C	Susceptible	[53]
Iranian	150/140	CC	Susceptible	[28]

Table 15.
Association of TGF-β1 (509 C/T) polymorphism with BD susceptibility.

12. Conclusion

In spite of recent advances in genetics and immunology leading to a better understanding of the immunopathogenesis, the etiology of BD is still unclear. Various genetic, immunological and micro- and macro-environmental factors are believed to be involved in the development of BD. The HLA-B*51 allele and variants in IL-10, TNF-α, TGF-β and at the IL-23–IL-12RB2 loci are the genetic factors most closely associated with BD. The variations in the association between various polymorphisms discussed and BD in different ethnicity/populations may reflect the heterogeneity in the genetic susceptibility to this disorder. Since the clear pathogenesis of BD remains to be elucidated, it is highly suggestive that multiple host genetic factors are involved in the development of BD. Therefore, further genetic studies on BD patients of different ethnicity and genes associated with immunity are expected to elucidate BD pathogenesis and also to contribute to the development of more targeted therapies and biomarkers.

Acknowledgements

Authors wish to thank MSD administration for facilities and support.

Conflict of interest

No conflicts of interests.

Author details

Abdulrahman Al Asmari* and Misbahul Arfin*
Medical Services Department of Armed Forces, Scientific Research Center, Riyadh, Saudi Arabia

*Address all correspondence to: abdulrahman.alasmari@gmail.com and misbahularfin@yahoo.com

References

[1] Yurdakul S, Yazici Y. Epidemiology of Behçet's syndrome and regional differences in disease expression. In: Yazici Y, Yazici H, editors. Behçet's Syndrome. 1st ed. New York: Springer; 2010. pp. 35-52

[2] Kapsimali VD, Kanakis MA, Vaiopoulos GA, Kaklamanis PG. Etiopathogenesis of Behcet's disease with emphasis on the role of immunological aberrations. Clinical Rheumatology. 2010;**29**:1211-1216. DOI: 10.1007/s10067-010-1491-6

[3] Davatchi F, Shahram F, Chams-Davatchi C, Shams H, Nadji A, et al. Behçet's disease in Iran: Analysis of 6500 cases. International Journal of Rheumatic Diseases. 2010;**13**:367-373. DOI: 10.1111/j.1756-185X.2010.01549.x

[4] Cho S, Cho S, Bang D. New insights in the clinical understanding of Behcet's disease. Yonsei Medical Journal. 2012;**53**:35-42. DOI: 10.3349/ymj.2012.53.1.35

[5] Pineton de Chambrun M, Wechsler B, Geri G, Cacoub P, Saadoun D. New insights into the pathogenesis of Behçet's disease. Autoimmunity Reviews. 2012;**11**(10):687-698. DOI: 10.1016/j.autrev. 2011. 11.026

[6] Evereklioglu C. Current concepts in the etiology and treatment of Behcet disease. Survey of Ophthalmology. 2005;**50**:297-350. DOI: 10.1016/j.survophthal.2005.04.009

[7] Direskeneli H, Saruhan-Direskeneli G. The role of heat shock proteins in Behcet's disease. Clinical and Experimental Rheumatology. 2003;**21**:S44-S48

[8] Stojanov S, Kastner D. Familial autoinflammatory diseases: Genetics, pathogenesis and treatment. Current Opinion in Rheumatology. 2005;**17**:586-599

[9] Chi W, Zhu X, Yang P, Liu X, Lin X, et al. Upregulated IL-23 and IL-17 in Behçet patients with active uveitis. Investigative Ophthalmology & Visual Science. 2008;**49**(7):3058-3064. DOI: 10.1167/iovs. 07-1390

[10] Jiang Z, Yang P, Hou S, Du L, Xie L, Zhou H, et al. IL-23R gene confers susceptibility to Behcet's disease in a Chinese Han population. Annals of the Rheumatic Diseases. 2010;**69**(7):1325-1328. DOI: 10.1136/ard.2009. 119420

[11] Mizuki N, Meguro A, Ota M, Ohno S, Shiota T, et al. Genome-wide association studies identify IL23R-IL12RB2 and IL10 as Behçet's disease susceptibility loci. Nature Genetics. 2010;**42**:703-706. DOI: 10.1038/ ng.624

[12] Sugita S, Kawazoe Y, Imai A, Kawaguchi T, Horie S, et al. Role of IL-22- and TNF-α-producing Th22 cells in uveitis patients with Behcet's disease. Journal of Immunology. 2013;**190**:5799-5808. DOI: 10.4049/ jimmunol.1202677

[13] Gheita TA, Gamal SM, Shaker I, El Fishawy HS, El Sisi R, Shaker OG, et al. Clinical significance of serum interleukin-23 and A/G gene (rs17375018) polymorphism in Behçets disease: Relation to neuro-Behçet, uveitis and disease activity. Joint Bone Spine. 2015;**82**(3):213-215. DOI: 10. 1016/j.jbspin.2014.10.008

[14] Al-Okaily F, Arfin M, Al-Rashidi S, Al-Balawi M, Al-Asmari A. Inflammation-related cytokine gene polymorphisms in Behçet's disease. Journal of Inflammation Research. 2015;**8**:173-180. DOI: 10.2147/ JIR. S89283

[15] Takeuchi M, Kastner DL, Remmers EF. The immunogenetics

of Behçet's disease: A comprehensive review. Journal of Autoimmunity. 2015;**64**:137-148. DOI: 10.1016/j.jaut.2015. 08.013

[16] Remmers EF, Cosan F, Kirino Y, Ombrello MJ, Abaci N, et al. Genome-wide association study identifies variants in the MHC class I, IL10, and IL23R-IL12RB2 regions associated with Behçet's disease. Nature Genetics. 2010;**42**:698-702. DOI: 10.1038/ng.625

[17] Gul A. Behcet's disease: An update on the pathogenesis. Clinical and Experimental Rheumatology. 2001;**19**:6-12

[18] Takeno M, Shimoyama Y, Kashiwakura J, Nagafuchi H, Sakane T, Suzuki N. Abnormal killer inhibitory receptor expression on natural killer cells in patients with Behcet's disease. Rheumatology International. 2004;**24**:212-216. DOI: 10.1007/s00296-003-0352-x

[19] Raziuddin S, Al-Dalaan A, Bahabri S, Siraj AK, Al-Sedairy S. Divergent cytokine production profile in Behçet's disease. Altered Th1/Th2 cell cytokine pattern. J Rheumatol. 1998;**25**:329-333

[20] Hamzaoui K, Hamzaoui A, Guemira F, Bessioud M, Hamza M, Ayed K. Cytokine profile in Behçet's disease patients. Relationship with disease activity. Scandinavian Journal of Rheumatology. 2002;**31**:205-210

[21] Ben Ahmed M, Houman H, Miled M, Dellagi K, Louzir H. Involvement of chemokines and Th1 cytokines in the pathogenesis of mucocutaneous lesions of Behçet's disease. Arthritis and Rheumatism. 2004;**50**:2291-2295. DOI: 10.1002/art.20334

[22] Akdeniz N, Esrefoglu M, Keles MS, Karakuzu A, Atasoy M. Serum Interleukin-2, Interleukin-6, tumor necrosis factor-alpha and nitric oxide levels in patients with Behcet's disease. Annals of the Academy of Medicine, Singapore. 2004;**33**:596-599

[23] Oztas MO, Onder M, Gurer M, Bukan AN, Sancak B. Serum interleukin 18 and tumour necrosis factor-levels are increased in Behcet's disease. Clinical and Experimental Dermatology. 2005;**30**:61-63. DOI: 10.1111/ j.13 65 -2230.2004.01684.x

[24] Akman A, Sallakci N, Coskun M, Bacanli A, Yavuzer U, et al. TNF-alpha gene 1031 T/C polymorphism in Turkish patients with Behçet's disease. Br J Dermatol. 2006;**155**(2):350-356. DOI: 10. 1111/j.1365-2133.2006.07348.x

[25] Turner DM, Williams DM, Sankaran D, Lazarus M, Sinnott PJ, Hutchinson IV. An investigation of polymorphism in the interleukin-10 gene promoter. European Journal of Immunogenetics. 1997;**24**:1-8

[26] Sadeghi A, Davatchi F, Shahram F, Karimimoghadam A, Alikhani M, et al. Serum profiles of cytokines in Behcet's disease. Journal of Clinical Medicine. 2017;**6**:49. DOI: 10.3390/ jcm6050049

[27] Shimizu J, Izumi T, Arimitsu N, Fujiwara N, Ueda Y, et al. Skewed TGFβ/Smad signalling pathway in T cells in patients with Behçet's disease. Clinical and Experimental Rheumatology. 2012;**30**(3 Suppl 72):S35-S39

[28] Shahram F, Nikoopour E, Rezaei N, Saeedfar K, Ziaei N, et al. Association of interleukin-2, interleukin-4 and transforming growth factor-beta gene polymorphisms with Behcet's disease. Clinical and Experimental Rheumatology. 2011;**29**(4 Suppl 67):S28-S31

[29] Pravica V, Perrey C, Stevens A, Lee JH, Hutchinson IV. A single nucleotide polymorphism in the first intron of the human IFN-gamma gene: Absolute correlation with a

polymorphic CA microsatellite marker of high IFN-gamma production. Human Immunology. 2000;**61**(9):863-866

[30] Bidwell J, Keen L, Gallagher G, Kimberly R, Huizinga T, et al. Cytokine gene polymorphism in human disease: On-line databases. Genes and Immunity. 1999;**1**(1):3-19. DOI: 10.1038/sj.gene.6363645

[31] Bidwell J, Keen L, Gallagher G, Kimberly R, Huizinga T, et al. Cytokine gene polymorphism in human disease: On-line databases, supplement 1. Genes and Immunity. 2001;**2**(2):61-70. DOI: 10.1038/sj.gene 636 3733

[32] Smith AJ, Humphries SE. Cytokine and cytokine receptor gene polymorphisms and their functionality. Cytokine & Growth Factor Reviews. 2009;**20**(1):43-59. DOI: 10.1016/j.cytogfr. 2008.11.006

[33] Park K, Kim N, Nam J, Bang D, Lee ES. Association of TNFA promoter region haplotype in Behcet's disease. Journal of Korean Medical Science. 2006;**21**:596-601. DOI: 10.3346/jkms.2006.21.4.596

[34] Liang Y, Xu WD, Zhang M, Qiu LJ, Ni J, et al. Meta-analysis of association between cytokine gene polymorphisms and Behcet's disease risk. International Journal of Rheumatic Diseases. 2013;**16**:616-624. DOI: 10.1111/1756-185X.12221

[35] Zhang M, Xu WD, Wen PF, Liang Y, Liu J, et al. Polymorphisms in the tumor necrosis factor gene and susceptibility to Behcet's disease: An updated meta-analysis. Molecular Vision. 2013;**19**:1913-1924

[36] Touma Z, Farra C, Hamdan A, Shamseddeen W, Uthman I, et al. TNF polymorphisms in patients with Behcet disease: A meta-analysis. Archives of Medical Research. 2010;**41**:142-146. DOI: 10.1016/j.arcmed. 2010. 02.002

[37] Song YW, Kang EH. Behçet's disease and genes within the major histocompatibility complex region. Modern Rheumatology. 2012;**22**:178-185. DOI: 10.1007/s10165-011-0542-4

[38] Ates A, Kinikli G, Duzgun N, Duman M. Lack of association of tumor necrosis factor-alpha gene polymorphisms with disease susceptibility and severity in Behcet's disease. Rheumatology International. 2006;**26**:348-353. DOI: 10.1007/s00296-005-0610-1

[39] Ben Ahmed M, Houman H, Ben Ghorbel I, Braham-Sfaxi A, Miled M, et al. Cytokine expression within mucocutaneous lesions of Behçet's disease: Involvement of proinflammatory and Th1 cytokines. Advances in Experimental Medicine and Biology. 2003;**528**:343-346

[40] Sharma S, Ghosh B, Sharma SK. Association of TNF polymorphisms with sarcoidosis, its prognosis and tumour necrosis factor(TNF)-alpha levels in Asian Indians. Clinical and Experimental Immunology. 2008;**151**:251-259. DOI: 10.1111/j.1365-2249.2007.03564.x

[41] Abraham LJ, Kroeger KM. Impact of the -308 TNF promoter polymorphism on the transcriptional regulation of the TNF gene: Relevance to disease. Journal of Leukocyte Biology. 1999;**66**:562-566. DOI: 10.1002/jlb. 66.4.562

[42] Wilson AG, Symons JA, McDowell TL, McDevitt HO, Duff GW. Effects of a polymorphism in the human tumor necrosis factor alpha promoter on transcriptional activation. Proceedings of the National Academy of Sciences of the United States of America. 1997;**94**:3195-3199. DOI: 10.1073/pnas.94.7.3195

[43] Abdallah AN, Cucchi-Mouillot P, Biteau N, Cassaigne A, Haras D, Iron A. Analysis of the polymorphism of the tumour necrosis factor (TNF)

gene and promoter and of circulating TNF-alpha levels in heart-transplant patients suffering or not suffering from severe rejection. European Journal of Immunogenetics. 1999;**26**:249-255

[44] Jeong P, Kim EJ, Kim EG, Byun SS, Kim CS, Kim WJ. Association of bladder tumors and GA genotype of −308 nucleotide in tumor necrosis factor-alpha promoter with greater tumor necrosis factor-alpha expression. Urology. 2004;**64**:1052-1056. DOI: 10.1016/j.urology.2004.06.018

[45] Ateş O, Dalyan L, Hatemi G, Hamuryudan V, Topal-Sarıkaya A. Analyses of functional IL10 and TNF-α genotypes in Behcet's syndrome. Molecular Biology Reports. 2010;**37**:3637-3641. DOI: 10.1007/s11033-010-0015-4

[46] Kamoun M, Chelbi H, Houman MH, Lacheb J, Hamzaoui K. Tumor necrosis factor gene polymorphisms in Tunisian patients with Behcet's disease. Human Immunology. 2007;**68**:201-205. DOI: 10.1016/j.humimm.2006.12.006

[47] Radouane A, Oudghiri M, Chakib A, Bennani S, Touitou I, Barat-Houari M. SNPs in the TNF-α gene promoter associated with Behcet's disease in Moroccan patients. Rheumatology (Oxford). 2012;**51**:1595-1599. DOI: 10.1093/rheumatology/kes141

[48] Amirzargar A, Shahram F, Nikoopour E, Rezaei N, Saeedfar K, et al. Proinflammatory cytokine gene polymorphisms in Behcet's disease. European Cytokine Network. 2010;**21**: 292-296. DOI: 10.1684/ecn.2009. 0209

[49] Lee EB, Kim JY, Lee YJ, Park MH, Song YW. TNF and TNF receptor polymorphisms in Korean Behcet's disease patients. Human Immunology. 2003;**64**:614-620

[50] Chang HK, Jang WC, Park SB, Nam YH, Lee SS, et al. The novel – G646A polymorphism of the TNF-a promoter is associated with the HLA-B51 allele in Korean patients with Behcet's disease. Scandinavian Journal of Rheumatology. 2007;**36**:216-221. DOI: 10.1080/03009740601154244

[51] Arayssi TK, Hamdan AR, Touma Z, Shamseddeen W, Uthman IW, et al. TNF polymorphisms in Lebanese patients with Behcet's disease. Clinical and Experimental Rheumatology. 2008;**26**:S130-S131

[52] Duymaz-Tozkir J, Gul A, Uyar FA, Ozbek U, Saruhan-Direskeneli G. Tumor necrosis factor-alpha gene promoter region -308 and -376 G>a polymorphisms in Behcet's disease. Clinical and Experimental Rheumatology. 2003;**21**:S15-S18

[53] Dilek K, Ozçimen AA, Saricaoğlu H, Saba D, Yücel A, et al. Cytokine gene polymorphisms in Behçet's disease and their association with clinical and laboratory findings. Clinical and Experimental Rheumatology. 2009;**27**(2 Suppl 53): S73-S78

[54] Bonyadi M, Jahanafrooz Z, Esmaeili M, Kolahi S, Khabazi A, et al. TNF gene polymorphisms in Iranian Azeri Turkish patients with Behcet's Disease. Rheumatol Int. 2009;**30**:285-289. DOI: 10.1007/s00296-009-1134-x

[55] Alayli G, Aydin F, Coban AY, Süllü Y, Cantürk F, et al. T helper 1 type cytokines polymorphisms: Association with susceptibility to Behcet's disease. Clinical Rheumatology. 2007;**26**:1299-1305. DOI: 10.1007/ s10067-006-0503-z

[56] Storz K, Löffler J, Koch S, Vonthein R, Zouboulis CC, et al. IL-6 receptor, IL-8 receptor and TNF alpha 238 (G/a) polymorphisms are not associated with Behcet's disease in patients of German or Turkish origin. Clinical and Experimental Rheumatology. 2008;**26**:S103-S106

[57] Abdolmohammadi R, Bonyadi M. Polymorphisms of promoter region of TNF-α gene in Iranian Azeri Turkish patients with Behçet's disease. Journal of Korean Medical Science. 2017;**32**(1): 33-37. DOI: 10.3346/ jkms.2017.32.1.33

[58] Akman A, Sallakci N, Kacaroglu H, Tosun O, Yavuzer U, et al. Relationship between periodontal findings and the TNF-alpha gene 1031T/C polymorphism in Turkish patients with Behçet's disease. Journal of the European Academy of Dermatology and Venereology. 2008;**22**(8):950-957. DOI: 10.1111/j.1468-3083.2008. 02678.x

[59] Lu LY, Cheng HH, Sung PK, Tai MH, et al. Tumor necrosis factor –beta +252A polymorphism is associated with systemic lupus erythematosus in Taiwan. Journal of the Formosan Medical Association. 2005;**104**:563-570

[60] Panoulas VF, NIkas SN, Smith JP, Douglas KM, Nightingale P, et al. Lymphotoxin 252 a > G polymorphism is common associates with myocardial infarction in patients with rheumatoid arthritis. Annals of the Rheumatic Diseases. 2008;**67**:1550-1556. DOI: 10.1136/ard.2007.082594

[61] Boraska V, Zeggini E, Groves CJ, Rayner NW, Skrabić V, et al. Family-based analysis of tumor necrosis factor and lymphotoxin-alpha tag polymorphism with type 1 diabetes in the population of South Croatia. Human Immunology. 2009;**70**:195-199. DOI: 10.1016/j.humimm.2008.12.010

[62] Al-Rayes H, Al-Swailem R, Albelawi M, Arfin M, Asmari A, Tariq M. TNF-α and TNF-β gene polymorphism in Saudi rheumatoid arthritis patients. Clin Med Insights Arthritis Musculoskelet Disord. 2011;**4**:55-63. DOI: 10.4137/ CMAMD.S6941

[63] Yamashita N, Kaneoka H, Kaneko S, Takeno M, Oneda K, et al. Role of gamma delta T lymphocytes in the development of Behcet's disease. Clinical and Experimental Immunology. 1997;**107**:241-247. DOI: 10.1111/j.1365-2249.1997.274-ce1159.x

[64] Triolo G, Accardo-Palumbo A, Dieli F, Ciccia F, Ferrante A, et al. Vgamma9/Vdelta2 T lymphocytes in Italian patients with Behçet's disease: Evidence for expansion, and tumour necrosis factor receptor II and interleukin-12 receptor beta1 expression in active disease. Arthritis Research & Therapy. 2003;**5**:R262-R268. DOI: 10.1186/ar785

[65] Messer G, Spengler U, Jung MC, Honold G, Blömer K, et al. Polymorphic structure of the tumor necrosis factor (TNF) locus: An Ncol polymorphism in the first intron of the human TNF-beta gene correlates with a variant amino acid in position 26 and a reduced level of TNF-beta production. The Journal of Experimental Medicine. 1991;**173**:209-219. DOI: 10.1084/jem.173.1.209

[66] Abraham LJ, French MAH, Dawkins RL. Polymorphic MHC ancestral haplotypes affect the activity of tumor necrosis factor alpha. Clinical and Experimental Immunology. 1993;**92**:14-18. DOI: 10.1111/j.1365-2249.1993.tb059 40.x

[67] Verity DH, Wallace GR, Vaughan RW, Kondeatis E, Madanat W, et al. HLA and tumour necrosis factor (TNF) polymorphisms in ocular Behcet's disease. Tissue Antigens. 1999;**54**:264-272

[68] Mizuki N, Inoko H, Sugmura K, Nishimura K, Nakamura S, et al. RFLP analysis in the TNF-beta gene and the susceptibility to alloreactive NK cells in Behcet's disease. Investigative Ophthalmology & Visual Science. 1992;**33**:3084-3090

[69] Akdis M, Aab A, Altunbulakli C, Azkur K, Costa RA, et al. Interleukins (from IL-1 to IL-38), interferons,

transforming growth factor β, and TNF-α: Receptors, functions, and roles in diseases. The Journal of Allergy and Clinical Immunology. 2016;**138**(4): 984-1010. DOI: 10.1016/j.jaci. 2016.06.033

[70] Fiorentino DF, Zlotnik A, Vieira P, Mosmann TR, Howard M, et al. IL-10 acts on the antigen-presenting cell to inhibit cytokine production by Th1 cells. Journal of Immunology. 1991;**146**:3444-3451

[71] Talaat RM, Ashour ME, Bassyouni IH, Raouf AA. Polymorphisms of interleukin 6 and interleukin 10 in Egyptian people with Behcet's disease. Immunobiology. 2014;**219**:573-582. DOI: 10.1016/j.imbio. 2014.03.004

[72] Wallace GR, Kondeatis E, Vaughan RW, Verity DH, Chen Y, et al. IL-10 genotype analysis in patients with Behcet's disease. Human Immunology. 2007;**68**:122-127. DOI: 10.1016/j. humimm.2006.11.010

[73] Khaib Dit Naib O, Aribi M, Idder A, Chiali A, Sairi H, et al. Association Analysis of IL10, TNF-α, and IL23R-IL12RB2 SNPs with Behçet's Disease Risk in Western Algeria. Front Immunol. 2013;**4**:342. DOI: 10.3389/ fimmu.2013.00342

[74] Wu Z, Zheng W, Xu J, Sun F, Chen H, et al. IL10 polymorphisms associated with Behçet's disease in Chinese Han. Human Immunology. 2014;**75**:271-276. DOI: 10.1016/j. humimm.2013.11.009

[75] Hu J, Hou S, Zhu X, Fang J, Zhou Y, et al. Interleukin-10 gene polymorphisms are associated with Behcet's disease but not with Vogt-Koyanagi-Harada syndrome in the Chinese Han population. Molecular Vision. 2015;**21**:589-603 eCollection 2015

[76] Afkari B, Babaloo Z, Dolati S, Khabazi A, Jadidi-Niaragh F, et al. Molecular analysis of interleukin-10 gene polymorphisms in patients with Behçet's disease. Immunology Letters. 2018;**194**:56-61. DOI: 10.1016/j. imlet.2017.12.008

[77] Montes-Cano MA, Conde-Jaldón M, García-Lozano JR, Ortiz-Fernández L, Ortego-Centeno N, et al. HLA and non-HLA genes in Behçet's disease: A multicentric study in the Spanish population. Arthritis Research & Therapy. 2013;**15**(5):R145. DOI: 10.1186/ ar4328

[78] Xavier JM, Shahram F, Davatchi F, Rosa A, Crespo J, et al. Association study of IL10 and IL23R-IL12RB2 in Iranian patients with Behçet's disease. Arthritis Rheum. 2012;**64**:2761-2772. DOI: 10.1002/art.34437

[79] Barış S, Akyürek Ö, Dursun A, Akyol M. The impact of the IL-1β, IL-1Ra, IL-2, IL-6 and IL-10 gene polymorphisms on the development of Behcet's disease and their association with the phenotype. Medicina Clínica (Barcelona). 2016;**146**(9):379-383. DOI: 10.1016/j.medcli.2015.09.017

[80] Karasneh J, Hajeer AH, Barrett J, Ollier WE, Thornhill M, Gul A. Association of specific interleukin 1 gene cluster polymorphisms with increased susceptibility for Behcet's disease. Rheumatology (Oxford). 2003;**42**:860-864. DOI: 10.1093/rheumatology/keg232R

[81] Coskun M, Bacanli A, Sallakci N, Alpsoy E, Yavuzer U, Yegin O. Specific interleukin-1 gene polymorphisms in Turkish patients with Behcet's disease. Exp Dermatol. 2005;**14**:124-129. DOI: 10.1111/j.0906-6705.2005.00253.x

[82] Ozcimen AA, Dilek K, Bingöl U, Sarıcaoğlu H, Sarandöl A, et al. IL-1 cluster gene polymorphisms in Turkish patients with Behcet's disease. International Journal of Immunogenetics. 2011;**38**:295-301. DOI: 10.1111/j.1744-313X.2011.01006.x

[83] Katayama T, Tachinami K, Ishiguro M, Kubota Y. The relation between Behçet's disease and interleukin-1 beta production. Nippon Ganka Gakkai Zasshi. 1994;**98**(2):197-201

[84] Fishman D, Faulds G, Jeffery R, Mohamed-Ali V, Yudkin JS, et al. The effect of novel polymorphisms in the interleukin-6 (IL-6) gene on IL-6 transcription and plasma IL-6 levels, and an association with systemic-onset juvenile chronic arthritis. The Journal of Clinical Investigation. 1998;**102**(7): 1369-1376. DOI: 10.1172/JCI2629

[85] Imani D, Rezaei R, Razi B, Alizadeh S, Mahmoudi M. Association between IL6-174 G/C polymorphism and Graves' disease: A systematic review and meta-analysis. Acta Medica Iranica. 2017;**55**(11):665-671

[86] Benešová Y, Vašků A, Bienertová-Vašků J. Association of interleukin 6, interleukin 7 receptor alpha, and interleukin 12B gene polymorphisms with multiple sclerosis. Acta Neurologica Belgica. 2018;**118**(3):493-501. DOI: 10.1007/s13760-018-0994-9

[87] Barartabar Z, Nikzamir A, Sirati-Sabet M, Aghamohammadi E, Chaleshi V, et al. The relationship between 174 G/C and −572 G/C of IL-6 gene polymorphisms and susceptibility of celiac disease in the Iranian population. Przegląd Gastroenterologiczny. 2018;**13**(4):293-298. DOI: 10.5114/pg.2018.79808

[88] Yousefi A, Najafi M, Motamed F, Mahmoudi E, Bidoki AZ, et al. Association of Interleukin-6 and Interleukin-1 family gene polymorphisms in autoimmune hepatitis. Annals of Hepatology. 2018;**17**(6):1021-1025. DOI: 10.5604/01.3001.0012.7202

[89] Yamakawa Y, Sugita Y, Nagatani T, Takahashi S, Yamakawa T, et al. Interleukin-6 (IL-6) in patients with Behçet's disease. Journal of Dermatological Science. 1996;**11**(3):189-195

[90] Chang HK, Jang WC, Park SB, Han SM, Nam YH, et al. Association between interleukin 6 gene polymorphisms and Behcet's disease in Korean people. Annals of the Rheumatic Diseases. 2005;**64**(2):339-340. DOI: 10.1136/ard.2004.024208

[91] Xu Y, Zhou K, Yang Z, Li F, Wang Z, Xu F, et al. Association of cytokine gene polymorphisms (IL 6, IL 12B, IL 18) with Behcet's disease: A meta-analysis. Z Rheumatol. 2016;**75**(9):932-938. DOI: 10.1007/s00393-015-0036-4

[92] Carapito R, Shahram F, Michel S, Le Gentil M, Radosavljevic M, et al. On the genetics of the silk route: Association analysis of HLA, IL10, and IL23R-IL12RB2 regions with Behçet's disease in an Iranian population. Immunogenetics. 2015;**67**(5-6):289-293. DOI: 10.1007/s00251-015-0841-6

[93] Kang EH, Kim S, Park MY, Choi JY, Choi IA, et al. Behçet's disease risk association fine-mapped on the IL23R-IL12RB2 intergenic region in Koreans. Arthritis Research & Therapy. 2017;**19**(1):227. DOI: 10.1186/s13075-017-1435-5

[94] Qin X, Xu J, Wu Z, Sun F, Chen H, et al. Association study of rs924080 and rs11209032 polymorphisms of IL23R-IL12RB2 in a Northern Chinese Han population with Behcet's disease. Human Immunology. 2016;**77**(12):1284-1290. DOI: 10.1016/j.humimm.2016.09.006

[95] Yu H, Zheng M, Zhang L, Li H, Zhu Y, et al. Identification of susceptibility SNPs in IL10 and IL23R-IL12RB2 for Behçet's disease in Han Chinese. The Journal of Allergy and Clinical Immunology. 2017;**139**(2):621-627. DOI: 10.1016/j.jaci.2016.05.024

[96] Li X, Bai L, Fang J, Hou S, Zhou Q, et al. Genetic variations of IL-12B, IL-12Rβ1, IL-12Rβ2 in Behcet's disease and VKH syndrome. PLoS One. 2014;**9**(5):e98373. DOI: 10.1371/journal.pone.0098373

[97] Jiang Z, Hennein L, Tao Y, Tao L. Interleukin-23 receptor gene polymorphism may enhance expression of the IL-23 receptor, IL-17, TNF-α and IL-6 in Behcet's disease. PLoS One. 2015;**10**(7):e0134632. DOI: 10.1371/journal.pone.0134632

[98] Yalçin B, Atakan N, Dogan S. Association of interleukin-23 receptor gene polymorphism with Behçet disease. Clinical and Experimental Dermatology. 2014;**39**(8):881-887. DOI: 10.1111/ced.12400

[99] Kirino Y, Bertsias G, Ishigatsubo Y, Mizuki N, Tugal-Tutkun I, et al. Genome-wide association analysis identifies new susceptibility loci for Behçet's disease and epistasis between HLA-B*51 and ERAP1. Nature Genetics. 2013;**45**(2):202-207. DOI: 10.1038/ng.2520

[100] Iwakura Y, Ishigame H. The IL-23/IL-17 axis in inflammation. The Journal of Clinical Investigation. 2006;**116**(5): 1218-1222. DOI: 10.1172/JCI28508

[101] Steinman L. Mixed results with modulation of TH-17 cells in human autoimmune diseases. Nature Immunology. 2010;**11**(1):41-44. DOI: 10.1038/ni.1803

[102] Chang JT, Shevach EM, Segal BM. Regulation of interleukin (IL)-12 receptor beta2 subunit expression by endogenous IL-12: A critical step in the differentiation of pathogenic autoreactive T cells. The Journal of Experimental Medicine. 1999;**189**(6):969-978. DOI: 10.1084/jem.189.6.969

[103] Kappen JH, Medina-Gomez C, van Hagen PM, Stolk L, Estrada K, et al. Genome-wide association study in an admixed case series reveals IL12A as a new candidate in Behçet disease. PLoS One. 2015;**10**(3):e0119085. DOI: 10.1371/journal. Pone.0119085

[104] Barksby HE, Lea SR, Preshaw PM, Taylor JJ. The expanding family of interleukin-1 cytokines and their role in destructive inflammatory disorders. Clinical and Experimental Immunology. 2007;**149**(2):217-225. DOI: 10.1111/j.1365-2249.2007.03441.x

[105] Schmitz J, Owyang A, Oldham E, Song Y, Murphy E, et al. IL-33, an interleukin-1-like cytokine that signals via the IL-1 receptor-related protein ST2 and induces T helper type 2-associated cytokines. Immunity. 2005;**23**(5):479-490. DOI: 10.1016/j.immuni.2005.09.015

[106] Chackerian AA, Oldham ER, Murphy EE, Schmitz J, Pflanz S, Kastelein RA. IL-1 receptor accessory protein and ST2 comprise the IL-33 receptor complex. Journal of Immunology. 2007;**179**(4):2551-2555. DOI: 10.4049/jimmunol.179.4.2551

[107] Palmer G, Lipsky BP, Smithgall MD, Meininger D, Siu S, et al. The IL-1 receptor accessory protein (AcP) is required for IL-33 signaling and soluble AcP enhances the ability of soluble ST2 to inhibit IL-33. Cytokine. 2008;**42**(3):358-364. DOI: 10.1016/j.cyto.2008.03.008

[108] Skef W, Hamilton MJ, Arayssi T. Gastrointestinal Behçet's disease: A review. World Journal of Gastroenterology. 2015;**21**(13):3801-3812. DOI: 10.3748/wjg.v21.i13.3801

[109] Koca SS, Kara M, Deniz F, Ozgen M, Demir CF, Ilhan N, et al. Serum IL-33 level and IL-33 gene polymorphisms in Behçet's disease. Rheumatology International. 2015;**35**(3):471-477. DOI: 10.1007/s00296-014-3111-2

[110] Wahl SM, Wong H, McCartney-Francis N. Role of growth factors in inflammation and repair. Journal of Cellular Biochemistry. 1989;**40**(2): 193-199. DOI: 10.1002/jcb.240400208

[111] Tsunawaki S, Sporn M, Ding A, Nathan C. Deactivation of macrophages by transforming growth factor-beta. Nature. 1988;**334**(6179):260-262. DOI: 10.1038/334260a0

[112] Broekelmann TJ, Limper AH, Colby TV, McDonald JA. Transforming growth factor beta 1 is present at sites of extracellular matrix gene expression in human pulmonary fibrosis. Proceedings of the National Academy of Sciences of the United States of America. 1991;**88**(15):6642-6646. DOI: 10.1073/pnas.88.15.6642

[113] Piga M, Mathieu A. Genetic susceptibility to Behcet's disease: Role of genes belonging to the MHC region. Rheumatology (Oxford, England). 2011;**50**:299-310. DOI: 10.1093/rheumatology/keq331

[114] El-Menyawi M, Fawzy M, Al-Nahas Z, Edris A, Hussein H, et al. Serum tumor necrosis factor alpha (TNF-α) level in patients with Behçet's disease: Relation to clinical manifestations and disease activity. The Egyptian Rheumatologist. 2014;**36**:139-143

[115] Balkwill F. TNF-alpha in promotion and progression of cancer. Cancer Metastasis Reviews. 2006;**25**:409-416. DOI: 10.1007/s10555-006-9005-3

[116] Glocker EO, Kotlarz D, Boztug K, Gertz EM, Schäffer AA, et al. Inflammatory bowel disease and mutations affecting the interleukin-10 receptor. The New England Journal of Medicine. 2009;**361**:2033-2045. DOI: 10.1056/NEJMoa0907206

[117] Guis S, Balandraud N, Bouvenot J, Auger I, Toussirot E, et al. Influence of –308A/G polymorphism in the tumor necrosis factor alpha gene on etanercept treatment in rheumatoid arthritis. Arthritis and Rheumatism. 2007;**57**:1426-1430. DOI: 10.1002/art.23092

[118] Seitz M, Wirthmuller U, Moller B, Villiger PM. The −308 tumor necrosis factor-α gene polymorphism predicts therapeutic response to TNFα-blockers in rheumatoid arthritis and spondyloarthritis patients. Rheumatology. 2007;**46**:93-96. DOI: 10.1093/rheumatology/kel175

Chapter 8

IL-21 Signaling and Induction of Cytokine Expression in Human Leukemia Cells and Monocytes

Chantel F. Faqua, Richard Akomeah and Samuel Evans Adunyah

Abstract

Interleukin-21 (IL-21) is produced by activated T cells and it plays many diverse roles by regulating the functions of normal and abnormal cells. Its roles include regulation of proliferation, promotion of immune system and activation of apoptosis in B cells. IL-21R is a type-1 cytokine receptor and belongs to the IL-2R and IL-15R family. The signaling mechanisms of IL-21 in different cell types have been identified. However, we know less about the biological effects of IL-21 and its signaling mechanisms in leukemia cells and monocytes. In this chapter, we will focus on IL-21's biological effects and signaling pathways as well as discuss the potential implications and applications of IL-21 in leukemia cells. In these cells, IL-21 does not promote proliferation but enhances apoptosis and chemotaxis. Furthermore, IL-21 promotes differential expression of many cytokines including interleukins and chemokines. IL-21 activates both the Raf-ERK-MAPK and the Jak/STAT signaling pathways. These pathways mediate some of the effects of IL-21. Lastly, IL-21 also promotes activation of the STAT3 promoter and other transcriptional factors. These findings may be relevant to IL-21's potential clinical implications and applications.

Keywords: IL-21, Leukemia, monocyte, ERK, Jak/STAT

1. Introduction

Interleukin-21 (IL-21) was first identified in 2000 through screening of a cDNA library from CD3+ human T cells [1]. Subsequently, the investigators identified a specific clone as containing pro-proliferative effects that was sensitive to neutralization by the soluble IL-21R. Later investigators characterized this clone as a factor, which was subsequently renamed IL-21 [2]. The sequence of this new cytokine gene has an open reading frame that encodes for a 163 amino acid peptide, later confirmed as IL-21. Matured IL-21 is 15 kDa and it has four helix bundle cytokine domains and two pairs of cysteine residues showing significant homology to IL-2, IL-4, IL-15 and GM-CSF [2–4]. Based on these structural characteristics, [1] IL-21 was placed in the IL-2 family. The IL-21 gene is located at chromosome 4q26–27 [1] and found to be approximately 180 kb from the IL-2 gene. In contrast, the human IL-15 gene is located on chromosome 4q31. Clearly, there is some commonality regarding the organization, gene structure, and location of these cytokines [1].

 IntechOpen

Human IL-21 has 62% homology with murine IL-21 in addition to their two pairs of cysteine residues and well conserved WSEWS motif [5] even though the murine IL-21 gene is located on chromosome 3 [1].

Initial report indicated that IL-21 is produced exclusively by CD4+ T cells during their response to stimulation by anti-CD3 and anti-CD8 antibodies [6]. However, other cell types are known to produce IL-21 [2–5]. IL-21 has effects on many cell types including human leukemia cells and human monocytes [2–7]. The IL-21 receptor (IL-21R) is expressed on both CD4+ and CD8+ T cells in which IL-21 stimulates proliferation [1]. In addition, IL-21 promotes allogeneic-specific proliferation of lymph node T cells, enhances cytolytic activity, induces IFN-γ production in T cells [2, 3] and enhances clonal expression of antigen-activated T cells [5, 6]. IL-21 also enhances naïve OT-1 cell response, enhances the affinity of the antigen-specific CD8+ T cells, and regulates the functions of both T cells and B-lymphocytes [1, 6, 7]. It also plays important roles in B cell proliferation, differentiation and class switching [8]. IL-21 causes induction of apoptosis of resting primary B cells via downregulation of anti-apoptotic factors, Bcl-2 and Bcl-xL [9] an effect which differentiates IL-21 from other members of the IL-21 family known to promote proliferative and survival [2, 3, 7, 8].

In B-chronic lymphocytic leukemia cells (B-CLL) IL-21 enhances production of granzyme B, which mediates B-CCL apoptosis when these cells were co-treated with IL-21 and CpG oligodeoxynucleotide [9]. Thus, IL-21 has anti-proliferative effects on B lymphoma cell line [3, 10, 11] suggesting that IL-21 has potential anti-tumor activity against B-cell malignancies. IL-21 enhances transcription of genes that encode IL-21R, and IL-18R, all of which are involved in innate immunity [10–13]. IL-21 expression is Ca^{2+}-dependent and that the IL-21 gene has three NFAT binding sites, which may be involved in regulation of the IL-21 gene promoter function by calcium ionophore [14]. In contrast to its known anti-proliferative effects [9, 10, 12, 13], IL-21 promotes growth, survival and induces DNA synthesis in human myeloma cell lines via Jak1/STAT2/3 mechanism [12]. In B cells, co-stimulation by IL-21 leads to three effects: growth arrest, apoptosis or growth [12–16]. Furthermore, IL-21 promotes maturation of NK cells [14], enhances differentiation of NK cells via co-stimulation with either IL-2, or IL-15 or IL-18 and regulates the functions of many different cell types [17–20].

IL-21 receptor (IL-21R) was first discovered in 2000 [1, 2] by investigators who identified a clone structure from the human chromosome 16p11 that led to their discovery of a protein of 538 amino acids [9]. This protein of 60 kDa was later identified as the type 1 receptor with about 62% homology to the IL-4 receptor (39 kDa) whose gene is located on chromosome 16 [1, 16]. The IL-21R has two subunits, alpha and gamma and it is expressed in B cells, T cells, dendritic cells, NK cells, myeloma cell lines, lymphoma cell lines (IM-9, NK-92 and Jarkat cells), and lymphoid tissues including spleen and thymus in both CD4+ and CD8+ T cells [17, 18]. IL-21R is upregulated upon T cell receptor activation [17, 18]. IL-21R is also expressed and upregulated in epidermis of systemic sclerosis patients [18]. IL-21R has two pairs of conserved cysteine residues in its extracellular domain and a WSXWS motif proximal to its transmembrane domain. In addition, it has two Box 1 and 2 motifs, which serve as docking sites for cytoplasmic Jak kinases involved in transducing cellular signaling from the IL-21R [21–25]. The γ-chain of IL-21R is an essential component of the IL-2 family of receptors that is critical for signaling as anti-γc antibodies blocked B cell proliferation induced by IL-21 and $CD4^0$, indicating that the gamma chain is involved in IL-21R signaling mechanisms [25, 26]. Furthermore, the Jak inhibitor WH-P131 blocked IL-21-induced proliferation of BaF3 cells via IL-21R confirming a role for Jak3 in IL-21R signaling. In addition, inhibition of the IL-21γc subunit by monoclonal antibody blocked IL-21-induced activation of Jak1, Jak3, and STAT1,

DOI: http://dx.doi.org/10.5772/intechopen.93004

STAT2, STAT3 and STAT5 thus further supporting a critical role for IL-21 γc subunit in IL-21 signaling [26].

IL-21R associates with BCL-6 translocation in B cell lymphomas and expression of IL-21R is upregulated by $CD4^0$ leading to transduction of pro-apoptotic signals [27, 28]. In addition, IL-21R plays important roles in immunoglobulin production and decreased in IL-21R expression is associated with decline in normal antibody production [28]. Furthermore, there is a decrease in IL-21R expression on B-lymphocytes in patients with systemic lupus erythromatosis (SLE) [29]. IL-21 synergizes with IL-15 to induce expansion of NK cells [1] and synergizes with IL-15 and IL-18 to upregulate IFNγ mRNA synthesis and production in NK and T cells. IL-21 also synergizes with IL-15 to promote proliferation of memory and naïve CD8+ [18], Furthermore, IL-21 has positive effect on tumor regression in B16 melanoma cells [24] as well as synergizes with IL-15 to delay CD11b bone marrow cell apoptosis [30]. In addition, IL-21 augments proliferation of IL-6-dependent human myeloma cells expressing IL-21R and synergizes with TNF to enhance myeloma cell mitogenesis [3, 13]. These synergistic effects of IL-21 led to consideration of IL-21 as a potential therapeutic agent for many cell malignancies [30]. In contrast to its synergetic effects, IL-21 antagonizes IL-4's ability to induce isotype switching in B cells from IgM to IgG and inhibits IFN-γ production by Th1 cells an [31, 32].

In BAB/C mice undergoing immunization, IL-21 caused reduction in antigen-induced eosinophil recruitment into the airway of the mice without affecting antigen-induced lymphocyte and macrophage recruitment [30]. However, another study [33] showed that IL-21 causes a decrease in antiovabumin IgE and IgG production, implying that IL-21 has anti-inflammatory effects in asthmatic patients and could serve as a potential therapy against asthma. The role of IL-21 in Crohn's disease (CD) has been reported [34]. Since Steven Rosenberg and his team [35] reported that IL-2 has anti-tumor in cancer patients, there has been extensive interest on whether members of IL-2 family of cytokines could generate similar effects. Subsequently, there are many reported that IL-21 alone or in combination with other cytokines produces anti-growth effect in various animal cancer models and on a variety of solid tumors including tumors without IL-21R [11, 36]. IL-21 is essential for IFN-γ induced expression of CXC chemokines, known to inhibit tumor angiogenesis. In addition, IL-21 induces tumor rejection by specific CTL and IFN-γ-dependent CXC chemokines in synergistic manner [37]. This anti-tumor effect of IL-21 is independent of the systemic cytokine release by known inflammatory mediators like IFN-y, IL-4, IL-10 and IL-12 [38]. Either alone or in combination with IL-23, IL-21 exhibits strong anti-tumor and anti-metastasis effects in kidney renal cell carcinoma in BALB/C mice and in human esophageal carcinoma tumors [39, 40]. Similarly, combination treatment with IL-15 and IL-21 promoted nearly 100% rejection of head and neck squamous cell carcinoma (HNSCC) in 30% of the experimental animals [41, 42], indicating potential role of IL-21 and IL-15 in minimizing transfection of animals with HNSCC [41, 42]. Similar reports in metastatic lymphoma animal models have indicate that IL-21 can act with other cytokines in combination therapy in the treatment of metastatic lymphoma tumors [43]. IL-21 enhances the synthesis of MIP-3α, a T cell chemoattractant and induce chemotaxis in intestinal epithelial cells (HT-29), indicating that IL-21 enhances immune cell response and could serve as an effective addition to cancer immunotherapy [43–45].

Our rationale for examining IL-21 effects in human myeloid monocytic leukemia cells and monocytes was based on the observations that IL-21 plays relevant roles in many types of cancer including solid tumors and lymphoma [36–43, 45]. However, little is known about the biologic effects of IL-21 in leukemia cells. In addition, while the signaling pathways of IL-21 in many cell types are known, not much is known about the signaling pathway utilized by IL-21 in leukemia cells and monocytes. Furthermore, knowledge on potential cytokine induction ability of IL-21 in leukemia cells and monocytes was lacking. Our earlier preliminary report indicates that STAT3, Smad1, Smad2, and Smad3 are regulated by IL-21 in U937 leukemia cells [46]. Additionally, IL-21 activates Jak/STAT and MAPK pathways in different cell types [47]. Therefore, the main purpose of our study was to determine the biological effects of IL-21 in human U937 leukemia and monocytes and to gain more insight into whether the Jak/STAT and MAPK-signaling pathways mediate IL-21's specific biologic effects in these cells.

2. Materials and methods

U937 myeloid monocytic leukemia cells were obtained from ATCC and cultured in RPMI-1640 media containing L-glutamine and supplemented with 10% heat-inactivated fetal bovine serum (FBS) (Atlanta Biological. GA) in the presence of 5 U/ml penicillin +50 U/ml of streptomycin (Invitrogen, CA). The cultures were maintained in 5% CO_2 at 37°C and 100% humidity. HeLa/STAT3-luc cells were created by co0transfection of pSTAT3-luc and pHygromycin into human cervical epithelial HeLa cells. These cells were maintained in the same culture in the presence of 100 μg/ml of hygromycin B (Roche, NJ) [7].

2.1 Monocyte isolation

Human PBMC were isolated from anonymous health individuals (NY Blood Center, Long Island, NY) by Ficoll gradient centrifugation at 400 × *g* for40 minutes at 20°C followed by Percoll gradient (GR Health Care, Piscataway, NJ) [7]. The mononuclear cells were recovered in RPMI media supplemented with 10% FBS, 100 μg of streptomycin per ml and 100 U of penicillin per ml. The white blood cells were layered onto plastic dishes and allowed to adhere at 37°C for 90 minutes to allow the suspended cells to adhere. At the end of the incubation period, the non-adherent cells were carefully discarded and the adherent monocytes were carefully removed and suspended in culture media. The monocytes were about 95% pure based on positive staining for CD14 marker.

2.2 Western blotting detection of proteins and phosphoproteins

Sixty million (6×10^6) U937 leukemia cells or monocytes were untreated (control) or treated with IL-21 (50 or 100 ng/ml) in a time course (2, 5, 15, 30, 60 minutes) experiments [48, 49]. At the end of the time course, the cells washed 2× with PBS and collected as pellet by centrifugation at 1800 rpm for 3 minutes. The cell pellets were washed two times with cold PBS and lysed in 500 μl of lysis buffer A (containing protease inhibitors, 0.5% Triton X-100, 50 mM NaF and 2 mM Vanadate) as described in [7]. Total cell lysate protein concentration was determined by Coomassie Blue Protein Assay Kit (Pierce, IL). Equal amount of lysate proteins (120 μg per sample) was resolved by 12% polyacrylamide gel electrophoresis followed by transfer of the proteins to membrane and blotted against specific antibodies to total Jak2 or Jak3 or Tyk2 or p-Jak2 or p-Jak3 or p-Tyk2 as previously described [7]. The immunoblot band intensities were scanned via spot densitometry analysis for quantitation and scanned intensity values from the IL-21 treated samples were compared to the intensity in the untreated samples.

2.3 Detection of IL-21R on U937 Leukemia cells and monocytes

Twenty million cells (20×10^6 cells) untreated or treated with 50 ng/ml of IL-21 for 24 hours were lysed in lysing buffer and protein the concentration was determined by Coomassie Blue protein determination assay kit [50]. Exactly, 200 μg/lane of total cell lysate was separated by 15% polyacrylamide gel electrophoresis and the proteins were subsequently transferred to nitrocellulose membrane. Following blocking of the background on the membrane, for detection of IL-21R, the nitrocellulose membranes containing transferred protein bands were incubated with 1:1000 dilution of IL-21 receptor primary antibody (Novus Biologicals) at 4°C in blocking buffer (TBS, 0.1% Tween 20, 5% nonfat dry milk) for 1 hour. Next, the membrane was washed 3x in water and incubated in milk blocking buffer containing 1:3000 dilution of goat anti-rabbit IgG HRP antibody (Amersham, CA) for 1 hour at room temperature. This was followed by washing with TBS buffer containing 0.5% Tween 20. Finally, the membrane was incubated for 1 minute with enhanced chemiluminescence (ECL) solution and exposed to Kodak film and band intensities were scanned via spot densitometry analysis.

2.4 Detection of Jak1 activation by Jak1 ELISA

To detect Jak1 by ELISA, 96 well plates were pre-coated with 10 μg/ml poly lysine for 30 minutes [51]. U937 cells were seeded at 20×10^3/well in 96 well plates and incubated in serum free medium overnight. Subsequently, the cells were either untreated or treated with 50 or 100 ng/ml of IL-21 for 2, 15, 30 or 60 minutes followed by rapid removal of the culture medium by aspiration. The cells in the wells were fixed with 100 μl of 8% formaldehyde in PBS for 20 minutes at room temperature. After removal of the formaldehyde all the wells were washed three times with 200 μl of wash buffer (0.1% Triton X-100 in PBS). After washing, 100 μl of quenching buffer (wash buffer with 1% hydrogen peroxide and 0.1% azide) was added to each well and incubated for 20 minutes at room temperature. Then, the wells were washed two times with wash buffer followed by incubation of the wells with 100 μl of antibody blocking buffer for 1 hour at room temperature. After removal of the antibody blocking buffer the wells were washed with 200 μl wash buffer followed by addition in 40 μl of diluted specific p-Jak1 or total Jak1 antibody. The wells were covered with parafilm, and incubated overnight at 4°C. The primary antibodies were removed by aspiration, and wells were washed three times with wash buffer. Next, 100 μl of diluted secondary antibody (goat-anti-rabbit HRP) was added to each well and incubated for 1 hour at room temperature with head-to-tail shaking. After the incubation, the secondary antibody was removed and wells were washed three times with wash buffer and once with PBS. The wells were aspirated to remove all traces of liquid, and 100 μl of developing solution was added to each well. The wells were incubated for 15 minutes at room temperature, while monitoring the color development. Thereafter, 100 μl of stop solution was added to each well. Lastly, absorbance was read at 450 nm using a microtiter plate reader. Experiments were conducted in triplicate.

2.5 STAT/DNA binding assays for detection of STAT activation by IL-21

To determine whetherIL-21 induces activation of STAT proteins, untreated and IL-21 treated U937 leukemia cells from time course experiments for specific STAT/DNA binding assays using the STAT transcription factor assay kits (Active Motif, Chemicon) [52, 53]. Following stimulation of cells with IL-21, we used the kit to monitor the activation or repression of several STAT proteins. The experiments were performed in triplicate.

2.6 Cell cycle analysis for cell growth

U937 cells (1×10^6) were either untreated or treated with 50 ng/ml of IL-21 for 24 or 48 hours [54, 55]. The cells were packed by centrifugation at 325 × *g* for 5 minutes, washed with cold PBS and suspended in 200 μl PBS. The cell suspension was mixed with 500 μl of ice cold ethanol and incubated on ice for 30 minutes. Next, the cells were harvested by centrifugation at 325 × *g* for 5 minutes and the pellets were suspended RNAse solution (containing 1 mg RNAse/ml in 0.2 M sodium phosphate buffer, pH 7.0) and 50 μl of propidium iodide (PI) solution (0.5 mg PI/ml in water). The samples were incubated at room temperature for 30 minutes. The tubes containing the cell suspensions were wrapped in aluminum foil and stored overnight at 4°C for equilibration. The cells were subjected to FACScan analysis using Cell Quest Pro (Becton Dickinson Immunocytometry system) and ModFitLT (v3.1, Versity Software House, Inc) to obtain G0, G1, S, and G2M population. For each analysis only single cells were analyzed. The data acquisition was set to collect a minimum of 104 cells within the single region established for each specimen using FL2-A vs. FL2-W plots. Data was analyzed using FL2-A. After establishing optimum conditions for data analysis, instrument settings were maintained constant throughout for all analysis.

2.7 Cell proliferation assays

To determine whether IL-21 stimulates cell proliferation in U937 cells, 6×10^6 cells were either untreated or stimulated with 100 ng/ml of IL-21 for up to 48 hours [56]. The cells were harvested at 24 or 48-hour time point. Aliquots of the harvested cells from each time point were diluted into 0.4% trypan blue/PBS solution at a ratio of 1:10. The cells were counted in triplicate and the average was recorded as the cell number for each sample. To validate the results of the trypan blue cell proliferation assay we performed MTT assay. Briefly, 6×10^6 cells were untreated or treated with 100 ng/ml of IL-21 for 24, 48 or 72 hours in 96 well plates. Next, we added 100 μl MTT (3-(4,5-dimethylthiazol-2-yl)-2,5-diphenyltetrazolium bromide) to each of the wells. The plates were incubated at 37°C to permit reduction of the MTT by the electrons from the viable cells. Four hours later, the MTT formazan produced in wells containing live cells appeared as black, fuzzy crystals at the bottom of the wells. Next, we added isopropanol and HCl (100 μl) to each of the wells and mixed thoroughly. After 1 hour, the absorbance of stable blue color developed in the upper phase of the cell pellets (indicative of cell viability) was measured in a microplate reader at 750 nm.

2.8 Caspase9 and Caspase3 activity assays as evidence for apoptosis

To determine whether IL-21 induces apoptosis in human U937 leukemia cells and human monocytes, we performed caspase9 and caspases3 enzymatic (colorimetric) assays in cell lysates from untreated and IL-21 treated cells. Leukemia cells (10×10^6) and human monocytes (10×10^6) were pretreated with orthovanade (5 mM) for 30 minutes [57–59]. The cells were either untreated or treated with 100 ng/ml of IL-21 for 24 or 48 hours. The cells were lysed in RIPA buffer, and total lysate protein concentration was determined as indicated above. To assess the lysate for caspases activities, 20 μg total lysate protein from each sample was added to each experimental assay well. For positive caspase activity, we used 20 μg of total cell lysates from U937 or monocytes treated with sodium butyrate, which is a strong inducer of apoptosis. Caspase9 and caspases3 activities were measured using specific assay kits (MBL International, Woburn, MA) using specific substrate (p-LEHD conjugate) for caspases9 and caspases3 (DEVD-pNA) respectively. The absorbance was read at 405 nm with microplate reader.

2.9 Chemotaxis assay

To determine whether culture medium from IL-21 causes chemotaxis, we placed culture media from untreated U937 cells or from U937 leukemia cells treated with IL-21 for 24 hours in the lower chamber of the Boden Chamber to serve as chemoattractant [60]. Human monocytes ((2×10^5) suspended in RPMI medium) were placed in the upper chamber. After 2-hour incubation in the Boyden Chamber, monocytes in both the lower and upper chambers counted as instructed by the kit supplier.

2.10 Luciferase assay

To determine whether IL-21 modulates Stat3 promoter function, we performed STAT3-Reporter gene analysis [61, 62]. STAT3-Reporter HeLa stable cells (Panomics, CA.) placed in 96 well plates for overnight incubation. The cells were either untreated or stimulated with different concentrations of IL-21 for different time points up to 24 hours. In some cases the cells were stimulated with different concentration of IL-21 (0.2, 5, 10, 12.5, 50, 75 or 100 μg/ml) for to 8–24 hours. As a positive control for the STAT3-Reporter gene activation, we treated some of the cells with Oncostatin M (Sigma, St. Louis) separately. At the end of the treatments, the medium was aspirated and the cells washed two times with PBS. The cells from each well were lysed in 100 μl lysing buffer and equal amount of protein in 20 μl of the lysate were added to 100 μl of Luciferase Assay Reagent (Promega) and the luciferin expression (as indicated by light production) was measured with a TD-20/20 luminometer.

3. Results

3.1 Effects of IL-21 on cell proliferation and apoptosis in both human leukemia cells and human monocytes

To determine whether IL-21 stimulates cell proliferation in myeloid monocytic leukemia cells, we performed proliferation assays on untreated and IL-21 treated leukemia cells. As shown in **Figure 1a**, IL-21 does not stimulate cell proliferation in human U937 myeloid monocytic leukemia. These cells express IL-21R. Therefore, it was surprising that IL-21 did not promote proliferation as indicated by the outcomes of both the trypan blue and MTT assays in cells treated with IL-21 for up to 48-hours. Furthermore, our results from cell cycle analysis (**Figure 1b**) show that IL-21 does not promote cell cycle progression during a 48-hour treatment. In contrast, IL-21 enhances accumulation of the leukemia cells at G0/G1 phase (**Figure 1c**), suggesting IL-21-induces apoptosis in the leukemia cells. This observation was validated by the results from specific caspase9 (**Figure 2a** and **b**) and caspases3 enzymatic assays (**Table 1**). As can be seen, IL-21stimulates several fold activation of both caspases9 and caspases3 activities in the IL-21 treated U937 cells. IL-21 produced far less apoptotic effect in monocytes as compared to the control (untreated) cells. The ability of IL-21 to stimulate apoptosis shown here is similar to an earlier reportin B cells [17].

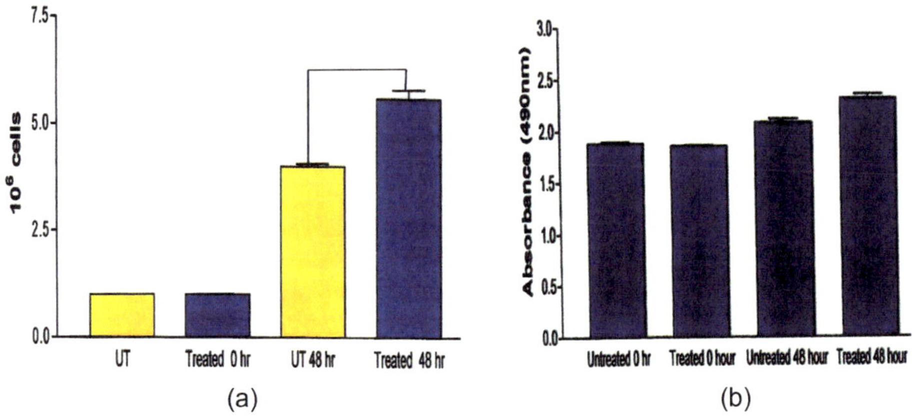

Figure 1.
(a and b) Effects of Il-21 on U937 leukemia cell growth.

3.2 Effects of IL-21 on cytokine and chemokine expression and ERK1/2 activity in human leukemia cells and monocytes

In our earlier reports [7, 46] we should that in both U937 leukemia cells and monocytes, IL-21 promoted time-dependent differential expression of several cytokines in the order of induction: IL7 > IL-15 > IL-2, IL-1, IFN-γ, TGFβ > GM-CSF and chemokines in the order: IL-8 > RANTES, IP-10 > MIP-1a > Eotaxin. Similar results were found in monocytes although to different extent. Furthermore, IL-21 triggers rapid phosphorylation of ERK-1/2 between 2 and 60 minutes in both leukemia cells and monocytes. IL-21-induced ERK1/2 phosphorylation is associated with ERK-1/2 enzymatic activation. The stimulatory effect of IL-21 on ERK-1/2 activation is significantly blocked by a neutralizing antibody against the IL-21R or impartially inhibited by the MEK inhibitor U0126 [7]. These results show that IL-21-induced interleukin and chemokine expression is partially mediated by ERK-1/2-dependent pathway.

3.3 IL-21 utilizes Jak-STAT signaling pathway

Next, we investigated whether IL-21 activates specific members of the Jak/Stat signaling pathway. Cell lysates from untreated cells or IL-21 stimulated cells were analyzed by Western Blot analysis using monoclonal antibody against specific members of the Jak/Stat signaling pathway. In some cases, we performed ELISA assay using specific ELISA assay kit (Ray Biotech Life, Peachtree Corner, Georgia, U.S.A) for specific members of the Jak/Stat signaling pathway [48, 49, 51, 63]. Our results (**Figure 3**) indicate IL-21 causes differential activation of the Jak kinases. Notably, Jak1 and Jak2 are rapidly activated in response to IL-21 stimulation; maximum stimulation occurring within 2–5 minutes. Even though Jak3 and Tyk2 are also activated in response to IL-21, the effect is delayed with maximum stimulation by IL-21 occurring at between 15 and 60 minutes. Because IL-21 stimulates significant activation of the Jak kinases, we examined whether specific members of the Stat family are also activated downstream of the Jak kinases. As shown (**Figure 4**), IL-21 differentially activates specific members of the STAT family. The order of activation was STAT2 > STAT4 > STAT6 > STAT3>. Clearly, IL-2 does not activate STAT1 and STAT5. In order to determine whether Jak2 and or Jak3 is responsible for activating Stat3 and Stat4, we examined the effect of Jak2 inhibitor and Jak3 inhibitor on STAT3 and STAT4 activation. Our results (**Figure 5**) show that in the presence of the

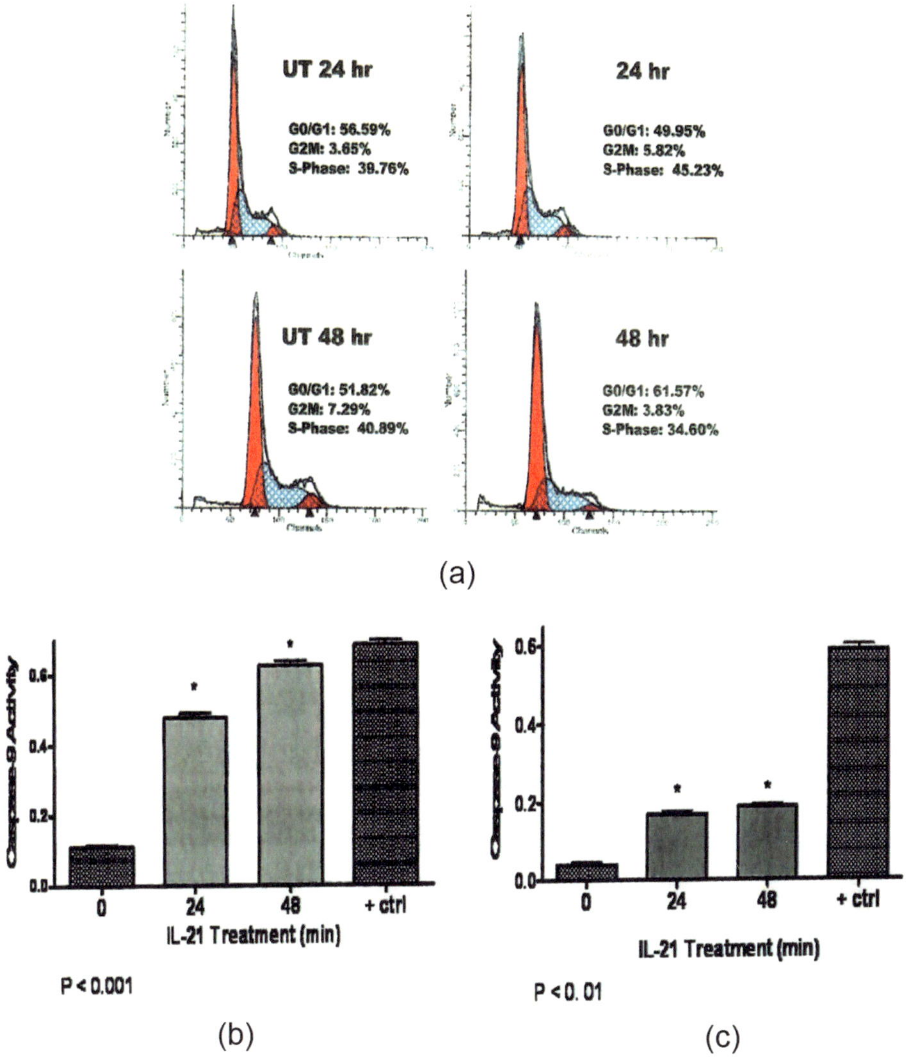

Figure 2.
(a) Effect of IL-21 on cell cycle progression. (b) IL-21-induced apoptosis in U937 leukemia cells. Asterisk (*) indicates significant difference between IL-21 treatment cells and untreated cells. (c) IL-21-induced apoptosis in monocytes. Asterisk (*) indicate significant difference between IL-21 treatment cells and untreated cells.

Experimental condition	U937 cells Caspase3 (fold activation)	Monocytes Caspase3 (fold activation)
Untreated	1	1
Cells + IL-21 (24 hours)	5.3	2.0
Cells + IL-21 (48 hours)	8.2	2.5

IL-21-Induced apoptosis in U937 cells and human monocytes. Leukemia cells (1×10^6) Or monocytes (1×10^6) were untreated (UT) or treated with 100 ng/ml of IL-21for 24 or 48 hours and cell lysate from each experiment was assayed for caspase3 activity. Data represents an average of two experiments.

Table 1.
IL-21-induced apoptosis in U937 leukemia cells and monocytes.

Jak2inhibitor, there is significant decline in both STAT3 activation. Clearly, our results show that both Jak2 and Jak3 are responsible for STAT3 activation [63]. In contrast, the Jak2 inhibitor and Jak3 inhibitor were only slightly effective in inhibiting STAT4

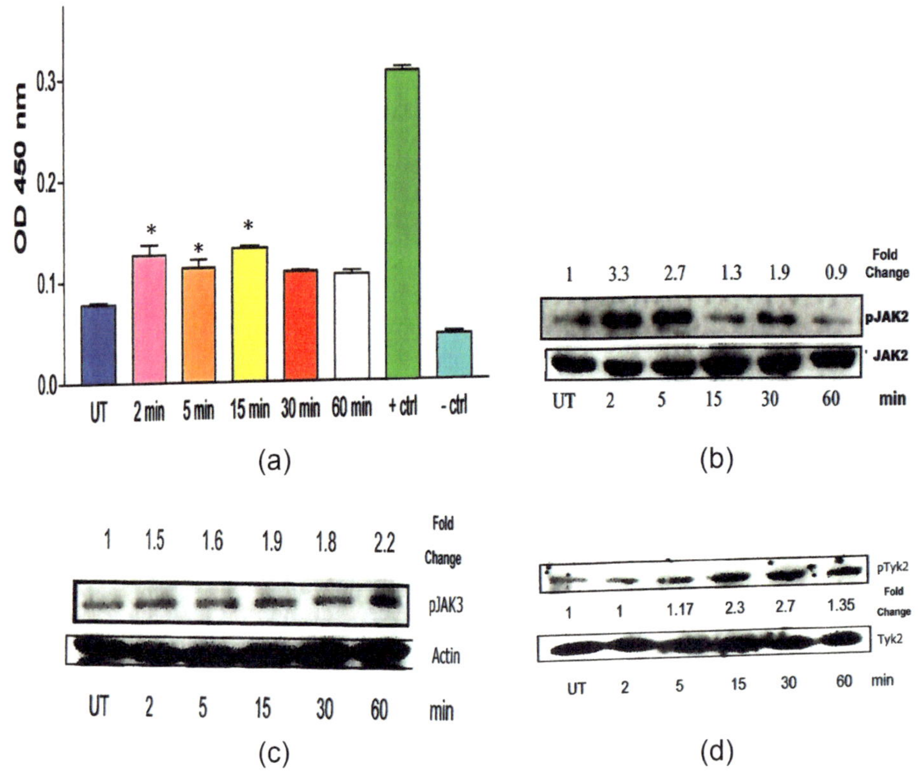

Figure 3.
(a) Effect of IL-21 on Jak1 activation. (b-d) Time course of IL-21-induced differential activation of Jak2, Jak3 and Tyk2.

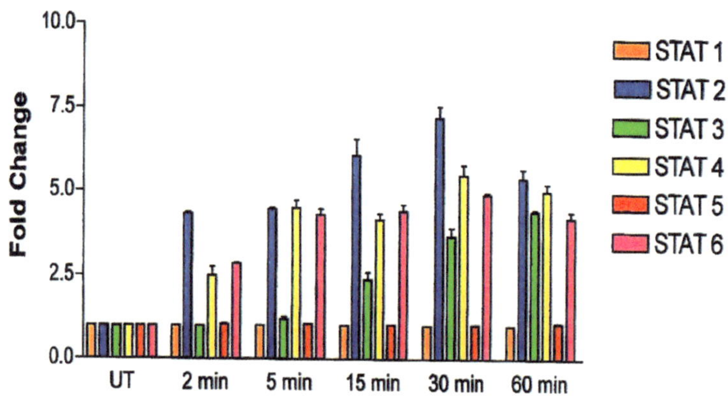

Figure 4.
IL-21 differentially activates STAT1, STAT2, STAT3, STAT4, STAT5 and STAt6.

activation. Hence, our results suggest Jak2 and Jak3may be involved in activating STAT3 [63]. These observations led us to conclude that, Jak2, Jak3 and STAT3 may be relevant for IL-21 induced signaling in leukemia cells and monocytes [63, 64].

3.3.1 Effect of IL-21 on chemotaxis: Role of ERK-1 and Jak/STAT proteins

In our previous report, we showed that IL-21 stimulates differential induction of cytokine and chemokine expression [7, 40] and that the effect is partially dependent

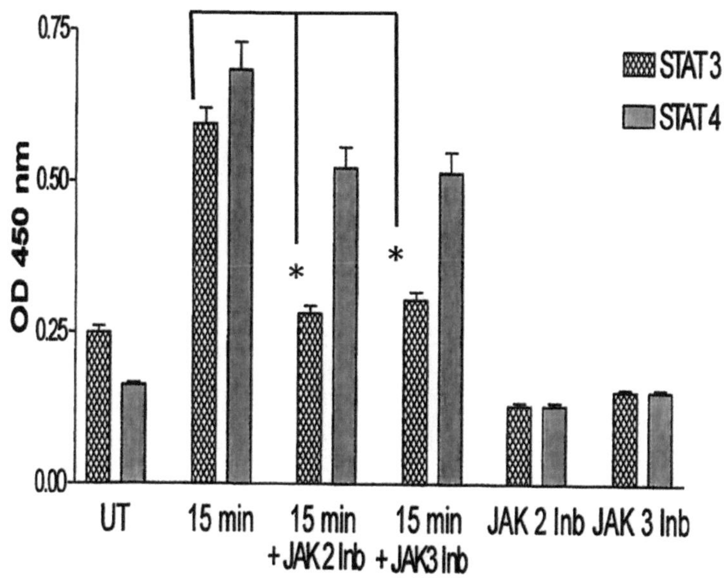

Figure 5.
Effect of Jak2 and Jak3 inhibitors on IL-21-induced STAT activation. Asterisk (*) indicates significant difference between IL-21 treatment alone and either IL-21 treatment in the presence of Jak2 inhibitor or Jak3 inhibitor.

on ERK1/2. Given that some of the chemokines including RANTES, MIP-1α could induce chemotaxis, we determined if the ERK-1/2 and Jak/Stat signaling pathways contributes to induction of chemotaxis. Therefore, we examined the effect of pre-incubation of cells with either ERK1/2 inhibitor or Jak2 inhibitor or with both prior to stimulation of the cells with IL-21 and assayed the cells for chemotaxis. The results (**Table 2**) confirmed that both the ERK1/2 and Jak2 contribute significantly towards IL-21-induced chemotaxis, suggesting that both pathways also contribute to IL-21-induced cytokine and chemokines expression. Thus, IL-21 utilizes both the ERK-1/2 MAPK and Jak/Stat signaling mechanisms for transducing its signal downstream the IL-21R. Perhaps, applications of anti ERK-1/2 and anti-Jak kinases agents could be relevant in blocking unwanted IL-21 effects in leukemia patients.

Experimental condition	Chemotaxis (fold change)
Untreated	1
Cells + IL-21	6.3 ± 0.2
Cells + U0126 + IL-21	3.5 ± 0.3
Cells + AG490 + IL-21	2.5 ± 0.1
Cells + U0126 + AG490 + IL-21	1.3 ± 0.1
Cells + U0126	0.9
Cells + Ag490	0.8

Evidence of involvement of ERK-1/2 and Jak2 in IL-21-induced chemotaxis. U937 leukemia cells were either untreated (UT) or treated with 100 ng/ml of IL-21 in the presence of either MEK inhibitor (U0126) or Jak2 inhibitor (AG490) or with both inhibitors. The effect of the each inhibitor alone was determined. Cells were incubated for 8 hours and culture medium from each experiment was placed in the bottom insert to serve as chemoattractant to monocytes (2 × 10^5) placed in the upper cup for Boyden Chamber chemotaxis assay. Data represents the mean plus SD from three experiments.

Table 2.
Evidence of involvement of ERK-1/2 and Jak2 in IL-21-induced chemotaxis.

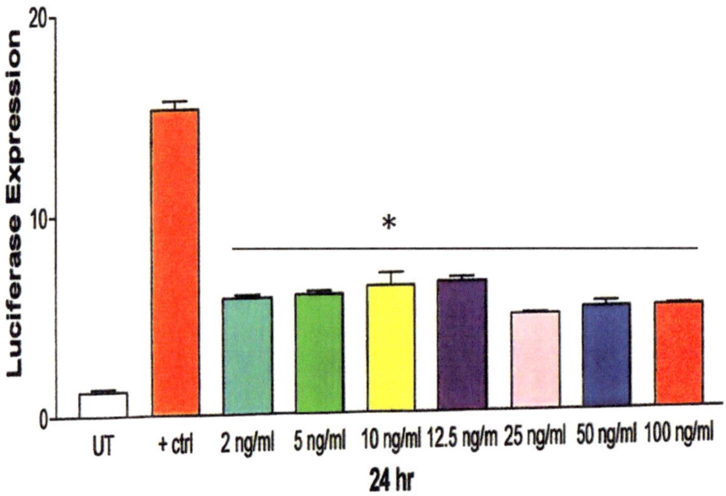

Figure 6.
Effect of IL-21 on STAT3 promoter cloned upstream of luciferase gene in STAT3-reporter-stable Hela cells. Asterisk (*) indicates significant difference between IL-21 treatment cells and untreated cells.

3.3.2 Effect of IL-21 on STAT3 promoter function

After showing that IL-21 regulates STAT3 phosphorylation and its DNA binding activity, we were curious whether IL-21 is capable of regulating the STAT3 promoter function. To address this question, we used a construct containing the STAT3 promoter cloned upstream the Luciferase gene (pSTAT3-Luc) in HeLa cells from human cervical cancer epithelial cells. We allowed cells to recover overnight and divided them into equal number of cells. The cells were either untreated or treated with different concentrations of IL-21 in time course experiments up to 24 hours. Luciferase assay was performed to determine the reporter gene function [61, 62]. The results in **Figure 6** indicate IL-21 induces a 2.5-fold increase in Luciferase activity in IL-21 treated cells as compared to untreated cells. From these results, we conclude that in addition to posttranslational regulation of STAT3, IL-21 is able to activate STAT3 promoter function. In addition, to STAT3 promoter regulation, we also noted that IL-21 regulates activation of several transcriptional factors including AP-2, Smad3 and 4, Pax-5, C-Myc, MEF-1 and CRE [46] that the effects of IL-21 were partially dependent on STAT3 activation since siRNA knockdown of STAT3 led to inhibition of IL-21's effects. By activating these transcriptional factors, IL-21 has broader regulatory roles that will likely impact on expression of genes downstream of these transcriptional factors with potential implicating consequences on the biology of different cell types.

4. Discussions

We have provided strong evidence that IL-21 does not promote proliferation in U937 myeloid monocytic leukemia cells. We have made nearly similar observations in THP-1 and HL-60 leukemia cell lines. However, IL-21 strongly promotes apoptosis in leukemia cells as has been reported for B cells [58]. In contrast, monocytes are somehow resistant to the apoptosis inducing effects of IL-21. These results are novel and could have interesting clinical implications and applications. Activation of caspases9 and caspases3 by IL-21 reported here, could point to involvement of either extrinsic or intrinsic apoptotic pathway or both and warrant further investigation. Currently, studies are underway to elucidate the potential involvement of mitochondrial in the mechanism of IL-21-induced apoptosis in these cells. Based on its ability to induce apoptosis in leukemia cells while sparing monocytes, IL-21 could be considered as potential anti-leukemia agent to kill myeloid monocytic leukemic cells [58, 59] or in combination with pro-apoptotic agents such as butyrate [65]. Likewise, a combination therapy of IL-21 with other cytokines such as IL-34, which promotes monocyte-like differentiation in leukemia cells, could be beneficial as anti-leukemia agent [66]. The stimulatory role of IL-21 in chemotaxis and invasion of cancer cells could be explored for development of anti-IL-21 drugs to combat IL-21-mediated invasive cancers as previously suggested [67–69].

In both leukemia cells and monocytes, IL-21 induces differential expression of IL-2, IL-7, IL-15, GM-CSF, TGF-β and IFNγ, which are known to regulate diverse cell functions. IL-15 is an effective immunotherapy in conjunction with IL-2. TGF-β induction by IL-21 is also very important since TGF-β is a regulator of cell growth and differentiation in many different cell types. GM-CSF is a promoter of colony formation during bone marrow hematopoietic differentiation as well as a stimulator of specific cell-mediated cytotoxicity against tumor targets. IFNγ and IL-7 are both relevant for regulation of cellular functions [7]. Thus, some of the effects can be beneficial in promoting immune defense in organisms in response to infection by pathogens.

The chemokines including IL-8, RANTES and MIP-1α, secreted to the microenvironment by leukemia cells and monocytes in response to IL-21 in leukemia could trigger of chemotaxis [7] as well as mediate cell–cell communication [70]. The inflammatory and pro-inflammatory cytokines and chemokines produced in response to IL-21 could enhance anti-body secretion and inflammatory responses [71, 72], as well as promote inflammatory diseases including arthritis and SLE [71–74]. Many of the cytokines induced by IL-21 in these cells could be relevant for immunotherapy. In effect, IL-21 could emerge as an effective adjunct immunotherapy for many types of cancer via T-cell activation, thus leading to increasing cytotoxic effects of NK cells [29]. However, we should not overlook the possibility that some of the cytokines and chemokines induced by IL-21 could alter the tumor microenvironment, enhance NK cell antitumor activity against MHC Class-I-deficient tumors, enhance protection against enteric microbial infections, mediate radiation-improved IL-21 cancer therapy, promote major oncological applications and triggering detrimental side effects including inflammation in cancer patients and Crohn's disease [74–80].

Our study was first to provide evidence that activation of the Raf-MEK-MAPK pathway by IL-21 is relevant for some of the biological effects of IL-21 in leukemia cells and monocytes [7]. As shown, IL-21 induces rapid activation of ERK1/2, an effect, which can either be blocked by a neutralizing antibody against the IL-21R or by the MEK inhibitor U0126. Furthermore, our study outcomes suggest that inhibition of ERK-1/2 leads to significant reduction of IL-21-induced cytokine and chemokine expression in leukemia cells and monocytes. However, inhibition of ERK-1/2 by the MEK inhibitor U0126 only partially abrogated IL-21-induced cytokine and chemokine expression, suggesting involvement of signaling pathway (s) independent of Raf-MEK-ERK-1/2.

The Jak/STAT signaling pathway plays major roles in the mechanisms of regulation of cellular functions. Therefore, we examined whether specific members of this pathway are activated by IL-21 in leukemia cells. Our results show that IL-21 activates Jak2, Jak 3 and Tyk2, which are involved in activating several STAT proteins including STAT2, STAT3, STAT4 and STAT6 in both U937 leukemia

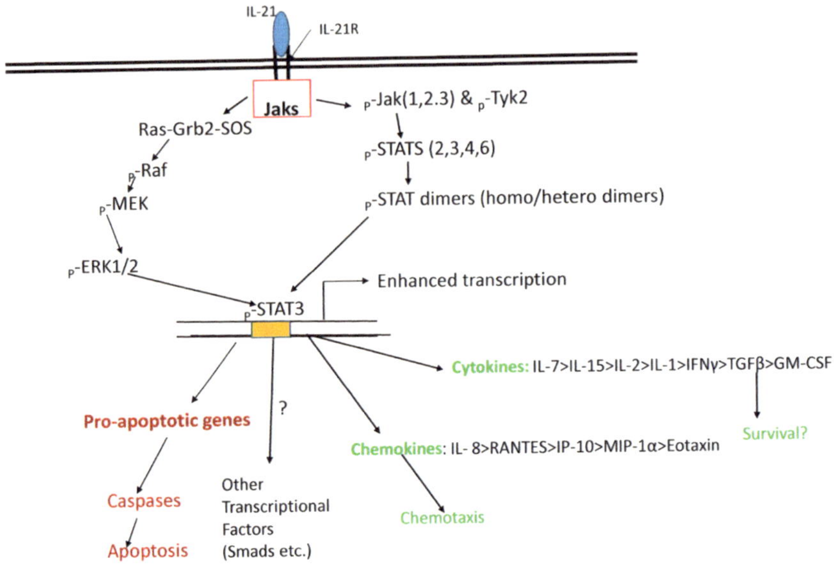

Figure 7.
Proposed signaling model for IL-21 in leukemia cells and monocytes.

cells and human monocytes. Using pharmacological inhibitors specific to either Jak2 or Jak3, we found that Jak2 and Jak3 play critical roles in mediating phosphorylation and activation of STAT3. Furthermore, the induction of chemotaxis by IL-21 is dependent on activation of both ERK-1/2 and Jak2. Our study also revealed that IL-21 activates Jak1 and Tyk2 as well as STAT1, STAT2 and STAT4 similar to earlier reports [71–75]. Taken together, both the Raf-MEM-ERK1/2 and Jak/STAT signaling pathways mediate IL-21-induced cytokine and chemokine expression in leukemia cells and monocytes. However, we are yet to elucidate the roles of STAT1, STAT2 and STAT4 in the biological effects of IL-21 in leukemia cells and monocytes. It remains possible that some of these STAT proteins activated by IL-21 in these cells will have opposing roles as reported in CD+ T cells [75].

The involvement of both the Raf-MEK-ERK and Jak/Stat signaling pathways in mediating signaling from the IL-21R and receptors of IL-2 and IL-15 is well documented [76, 77]. However, our results were among the first to clearly implicate both of these two signaling pathways in the mechanism by which IL-21 induces expression of cytokine and chemokines in leukemia cells and normal monocytes. These observations are interesting in view of the fact that both of these two signaling pathways support cell survival and proliferation and yet IL-21 fails to promote proliferation in the leukemia cells. Additionally, we have shown that IL-21 activates the STAT3 promoter suggesting that genes including anti-apoptotic and pro-apoptotic genes with STAT3 promoter can be regulated by IL-21 in leukemia cells. In addition, IL-21 regulates other transcriptional factors including Smad3/4, AP-2, Pax-5, C-myc, MEF-1 and CRE via partial STAT3-dependent mechanism. As illustrated by our model (**Figure 7**) upon binding to its receptor, IL-21 activates multiple signaling pathways, which are critical to the observed IL-21's biological effects in leukemia cells and monocytes.

The potential for IL-21 to play a role in immunotherapy including its role in cancer therapy is extensively documented [77–82]. Our work points to the need to expand research on IL-21 in different leukemia cells with the anticipation that the outcomes will provide new ideas for exploring IL-21 in combination therapy for leukemia. Lastly, we propose that targeting the major pathways modulated by IL-21 could lead to new opportunities for treating leukemia.

5. Conclusion

Our findings indicate that both U937 leukemia cells and human monocytes express IL-21R. In these cells, IL-21 induces differential expression of various cytokines and chemokines and promotes chemotaxis. These biological effects of IL-21 could be associated with either negative clinical implications or positive clinical applications in leukemia patients. The Raf-MEM-ERK1/2 pathway and the Jak-STAT pathway play critical roles in induction of cytokine and chemokine expression by IL-21. Activation of the Jak2 and STAT3 is germane to these effects. Furthermore, IL-21 activates apoptosis in U937 leukemia cells with little or no apoptotic effect on human monocytes. The inability of IL-21 to induce apoptosis in human monocytes while inducing significant apoptosis in leukemia cells could form the basis for future application of IL-21 as a potential therapeutic factor for various types of leukemia. Lastly, the ability of IL-21 to regulate the STAT3 promoter function suggests that genes including pro-apoptotic or anti-apoptotic genes with STAT3 promoter sequences are likely to be modulated by IL-21.

Acknowledgements

This work was supported by research funds from NIGMS/NIH SCORE, 5T32HL007735, Meharry-VICC Cancer Partnership NCI U54 grant (U54 CA091408) and current Meharry-VICC-TSU NCI U54 grant (5U54CA163069) to Professor Samuel E. Adunyah (SEA). In addition, SEA was supported by NIMHD MeTRC U54 grant (U54MD007593-09) during the preparation of this manuscript. Fuqua was supported by 5T32HL007735 from NHLBI/NIH. We are also grateful to Dr. J. O Price of Vanderbilt VA for flow cytometry work and to Ms. Ketia Barnes and Mrs. Marian Asmah-Addai for Admin support.

Author details

Chantel F. Faqua, Richard Akomeah and Samuel Evans Adunyah*
Department of Biochemistry, Cancer Biology, Neuroscience and Pharmacology, Meharry Medical College, Nashville, TN, USA

*Address all correspondence to: sadunyah@mmc.edu

References

[1] Parrish-Novak J, Dillion JR, Nelson A, Hammond A, Sprecher C, Gross JA, et al. IL-21 and its receptor are involved in NK cell expansion and regulation of lymphocyte function. Nature. 2000;**408**(6808):57-61

[2] Parrish-Novak J, Foster DC, Holly RD, Clegg CHJ. Interleukin-21 and the IL-21 receptor: Novel effectors of NK and T cell responses. Journal of Leukocyte Biology. 2002;**72**:856-863

[3] Brenne AT, Ro TB, Waage A, Sundan A, Borset M, Hjorth-Hansen H. Interleukin-21 is a growth and survival factor for human myeloma cells. Blood. 2002;**99**(10):3756-3762

[4] Sivakumar PV, Foster DC, Clegg CH. Interleukin-21 is a T-helper cytokine that regulates humoral immunity and cell-mediated anti-tumor responses. Immunology. 2004;**112**(2):177-182

[5] Ozaki K, Kikly K, Michalovich D, Young PR, Leonard WJ. Cloning of a type I cytokine receptor most related to the IL-2 receptor beta chain. Proceedings of the National Academy of Sciences of the United States of America. 2000;**97**:1439-1444

[6] Kuchen S, Robbins R, Sims GP, Sheng C, Phillips TM, Lipsky PE, et al. Essential role of IL-21 in B cell activation, expansion, and plasma cell generation during CD4$^+$ T cell-B cell collaboration. Journal of Immunology. 2007;**179**(9):5886-5896

[7] Fuqua CF, Akomeah R, Price JO, Adunyah SE. Evidence for the involvement of ERK1/2 in IL-21-induced cytokine production in leukemia cells and human monocytes. Cytokine. 2008;**44**:101-107

[8] Wan CK, Oh J, Li P, West EE, Wong EA, Andraski AB, et al. The cytokines IL-21 and GM-CSF have opposing regulatory roles in the apoptosis of conventional dendritic cells. Immunity. 2013;**38**(3):514-527. DOI: 10.1016/j.immuni.2013.02.011

[9] Weiner GJ, Taylor CM, Jacobs LS, Andreski M, Huang J, Woolridge JE, et al. B-chronic lymphocytic leukemia cells and other B cells can produce granzyme B and gain cytotoxic potential after Interleukin-21-based activation. Blood. 2006;**108**(8):2712-2719

[10] Webb R, Merrill JT, Kelly JK, Sestak A, Kaufman KM, Langefeld CD, et al. A polymorphism within interleukin-21 receptor (*IL21R*) confers risk for systemic lupus erythematosus. Arthritis and Rheumatism. 2009;**60**(8):2402-2407. DOI: 10.1002/art.24658

[11] Kasaian MT, Whitters MJ, Carter LL, Lowe LD, Jussif JM, Deng B, et al. IL-21 limits NK cell response and promotes antigen specific T cell activation: A mediator of transition from innate to adaptive immunity. Immunity. 2000;**16**(4):559-569

[12] Strengell M, Sareneva T, Foster D, Julkunen I, Matikainen S. IL-21 up-regulates the expression of genes associated with innate immunity and Th1 response. Journal of Immunology. 2002;**169**(7):3600-3605

[13] Moroz A, Eppolito PD, Nelson A, Clegg CH, Shrikant PA. IL-21 enhances and sustains CD$^+$8 T cell responses to achieve durable tumor immunity: Comparative evaluation of IL-2, IL-15 and IL-21. Journal of Immunology. 2004;**173**:900-9008

[14] Kim HP, Korn LL, Gamero AM, Leonard WJ. Calcium-dependent activation of IL-21 gene expression in T cells. The Journal of Biological Chemistry. 2005;**280**(16):25291-25297

[15] Ozaki K, Spolski R, Ettinger R, Kim HP, Wang G, Qi CF, et al. Regulation of B cell differentiation and plasma cell generation by IL-21, a novel inducer of Blimp-1 and Bcl-6. Journal of Immunology. 2004;**173**(9):5361-5371

[16] Ettinger R, Sims GP, Fairhurst AM, Robbins R, Sing da Silva Y, Spolski R, et al. IL-21 induces differentiation of human naïve and memory B cells into antibody-secreting plasma cells. Journal of Immunology. 2005;**175**(12):7867-7879

[17] Mehta DS, Wurster AL, Whitters MJ, Young DA, Collins MJ, Grusby MJ. IL-21 induces the apoptosis of resting and activated primary B cells. Journal of Immunology. 2003;**170**(8):4111-4118

[18] Nelson A, Hughes SD, Holly R, Heipel M, Johnson B, Bannink K. Anti-tumor effects of IL-21. American Science of Hematology Annual Conference. Blood. 2002;**100**:593

[19] Jin H, Carrio R, Yu A, Malek TR. Distinct activation signals determine whether IL-21 induces B cell costimulation, growth or Bim-dependent apoptosis. Journal of Immunology. 2004;**173**(1):657-665

[20] Sivori S, Cantoni C, Parolini S. IL-21 induces both rapid maturation of human CD34+ cell precursors towards NK cells and acquisition of surface killer Ig-like receptors. European Journal of Immunology. 2003;**33**(12):3439-3447

[21] Brady J, Hayakawa Y, Smyth MJ, Nutt SL. IL-21 induces the functional maturation of murine NK cells. J. Immunology. 2004;**172**(4):2048-2058

[22] Ozaki K, Kikly K, Michalovich D, Young PR, Leonard WJ. Cloning of a type I cytokine receptor most related to the IL-21 receptor beta chain. Proceedings of the National Academy of Sciences of the United States of America. 2000;**97**(21):11439-11444

[23] Collins M, Whitters MJ, Young DA. IL-21 and IL-21 receptor: A new cytokine pathway modulates innate and adaptive immunity. Journal of Immunology Research. 2003;**28**(2):131-140

[24] Distler JH, Jüngel A, Kowal-Bielecka O, Michael BA, Gay RE, Sprott H, et al. Expression of IL-21R in epidermis from patients with systemic sclerosis. Arthritis and Rheumatism. 2005;**52**(3):856-864

[25] Habib T, Senadheera S, Weinberg K, Kaushansky K. The common gamma chain is a required signaling component of the IL-21 receptor and supports IL-21 induced cell proliferation via Jak3. Biochemistry. 2002;**41**(27):8725-8731

[26] Asao H, Okuyama C, Kumaki S, Ishii N, Tsuchiya S, Foster D, et al. Cutting edge: The common gamma-chain is an indispensable subunit of IL-21 receptor complex. Journal of Immunology. 2001;**167**(1):1-5

[27] Udea C, Akasaka T, Kurata M, Maesako Y, Nishikori M, Ichinohasma R, et al. The gene for IL-21 receptor is the partner of BcL6 in t(3;16)(q27;p11), which is recurrently observed in diffuse large B-cell lymphoma. Oncogene. 2002;**21**(3):368-376

[28] Ferrini S, Ferrarini M, Gobbi M, Azzarone B, Capaia M, Fabbi M, et al. IL-21R is upregulated by CD40 triggering and mediates proapoptotic signals in chronic lymphocytic leukemia B cells. Blood. 2006;**107**(9):3708-3715

[29] Mitoma H, Horiuchi T, Kimoto Y, Tsukamoto H, Uchino A, Tamimoto Y, et al. Decreased expression of IL-21 receptor on peripheral B lymphocytes in systemic lupus erythematosus. International Journal of Molecular Medicine. 2005;**16**(4):609-615

[30] Zeng R, Spolski R, Finkelstein S, Oh S, Kovanen P, Hinriches C, et al.

Synergy of IL-21 with IL-15 in regulating CD8+ T cell expansion and function. The Journal of Experimental Medicine. 2005;**201**(1):139-148

[31] Pelletier M, Girard D. Differential effects of IL-15 and IL-21 in myeloid (CD11b+) and lymphoid (CD11b-) bone marrow cells. Journal of Immunology. 2006;**177**:100-108

[32] Suto A, Nakajima H, Hirose K, Suzuki K, Kagami SI, Seto Y, et al. IL-21 prevents antigen-induced IgE production by inhibiting germ line C transcription of IL-4 stimulated B cells. Blood. 2002;**100**(13):4565-4573

[33] Suto A, Wurster A, Reiner S, Grusby M. IL-21 inhibits IFN-γ production of developing Th1 cells through the repression of eomesodermin expression. Journal of Immunology. 2006;**177**(6):3721-3727

[34] Monteleone G, Monteleone I, Fina D, Vavassori P, Del Vecchio Blanco GL, Caruso R, et al. IL-21 enhances T-helper cell type 1 signaling and IFN-γ production in Crohn's disease. Gastroenterologia. 2005;**128**(3):687-694

[35] Rosenberg SA, Mule JJ, Spiess PJ, Reichert CM, Schwarz SL. Regression of established pulmonary metastases and subcutaneous tumor mediated by the systemic administration of high doses recombinant IL-2. The Journal of Experimental Medicine. 1985;**161**(5):1169-1188

[36] Wang G, Tschoi M, Spolski R, Yanyan L, Ozaki K, Feng C, et al. In vivo anti-tumor activity of IL-21 mediated by natural killer cells. Cancer Research. 2003;**63**:9016-9022

[37] Di Carlo E, Comes A, Orengo AM, Rosso O, Meazza R, Musiani P. IL-21 induces tumor rejection by specific CTL and IFN-γ-dependent CXC chemokines in syngeneic mice. The Journal of Immunology. 2004;**172**(3):1540-1547

[38] Ma HL, Whitters MJ, Konz RF, Senices M, Young DA, Grusby MJ, et al. IL-21 activates both innate and adaptive immunity to generate potent anti-tumor responses that requires perforin but are independent of IFN-gamma. Journal of Immunology. 2003;**17**(2):608-615

[39] Hughes S, Chin L, Waggie K, Chris C. Interleukin-21 efficacy in a mouse model of metastatic renal cell carcinoma. Journal of Clinical Oncology. 2004;**22**(14 Suppl):2598-2598

[40] Shan B, Yu L, Li Q, Shimozato O, Li Q, Tagawa M. Expression of IL-21 and IL-23 in human esophageal tumors produce anti-tumor effects in nude mice. Anticancer Research. 2004;**24**(1):79-82

[41] Nakano H, Kishda T, Asada H, Shin-Ya M, Shinomiya T, Imanishi J, et al. Interleuken-21 triggers both cellular and humoral immune responses leading to therapeutic anti-tumor effects against head and neck squamous cell carcinoma. J. Genetics in Medicine. 2006;**8**(1):90-99

[42] Ugai S, Shimozato O, Wang YQ, Kawamura K, Yamamoto H, Yamaguchi T, et al. Expression of the Interleukin-21 gene in murine colon carcinoma cells generates systemic immunity in the inoculated hosts. Cancer Gene Therapy. 2003;**10**(3):187-192

[43] Kishida T, Asada H, Itokawa Y, Cui FD, Shin-Ya M, Gojo S, et al. IL-21 and IL-15 genetic transfer synergistically augments therapeutic anti-tumor immunity and promotes regression of metastatic lymphoma. Molecular Therapy. 2003;**8**(4):552-558

[44] Caruso R, Fina D, Peluso I, Stolfi C, Fantini MC, Gioia V, et al. A functional role for interleukin-21 in promoting the synthesis of the T cell chemoattractant MIP-3α, by gut epithelial cells. Gastroenterologia. 2007;**132**(1):166-175

[45] Spolski R, Leonard WJ. Interleukin-21: Basic biology and implications for cancer and autoimmunity. Annual Review of Immunology. 2008;**26**:57-79

[46] Fuqua CF, Akomeah R, Adunyah SE. IL-21 activates key transcriptional factors including Stat3/5 and signal transducers Smad1/2/3 in U937 leukemia cells. In: Proceedings of 6th International Cytokine Conference (Editor: Josef D. Schwarzmeier); 27-31 August 2006. Vienna (Austria). pp. 61-66

[47] Davis ID, Skak K, Smyth MJ, Kristjansen PEG, Miller DM, Sivakumar PV. Interkeukin-21 signaling: Functions in cancer and autoimmunity. Clinical Cancer Research. 2007;**13**(23):6926-6931

[48] Subramanian VS, Pearson LL, Adunyah SE. Interleukin-17 induces rapid tyrosine phosphorylation and activation of Raf-1 kinase in human monocytic progenitor cell line U937. Biochemical and Biophysical Research Communications. 1999;**259**(1):172-177

[49] Subramanian VS, Cooper RS, Adunyah SE. Evidence for the involvement of JAK/STAT pathway in the signaling mechanism of Interleukin-17. Biochemical and Biophysical Research Communications. 1999;**262**(1):14-19

[50] deTotero TD, Meazza R, Simona Zupo S, Cutrona G, Matis S, Colombo M, et al. Interleukin-21 receptor (IL-21R) is up-regulated by CD40 triggering and mediates proapoptotic signals in chronic lymphocytic leukemia B cells. Blood. 2006;**107**:3708-3715

[51] Olorunseun OO, Beales ILP. The anti-apoptotic and growth stimulatory actions of leptin in human colon cancer cells involves activation of JNK mitogen activated protein kinase, JAK2 and PI3 kinase/Akt. International Journal of Colorectal Disease. 2007;**22**:401-409

[52] Lafarge S, Hamzeh-Cognasse H, Chavarin P, Genin C, Garraud O, Cognasse F. A flow cytometry technique to study intracellular signals NF-kappaB and STAT3 in peripheral blood mononuclear cells. BMC Molecular Biology. 2007;**8**(64):1-9

[53] Jiang H, Yu J, Guo H, Song H, Chen S. Upregulation of survivin by leptin/STAT3 signaling in MCF-7 cells. Biochemical and Biophysical Research Communications. 2008;**368**(1):1-5

[54] Wang W, Heideman L, Chung CS, Pelling JC, Koehler KJ, Birt DF. Cell-cycle arrest at G2/M and growth inhibition by apigeninin human colon carcinoma cell lines. Molecular Carcinogenesis. 2000;**28**:102-110

[55] Gray JW, Dolbeare F, Pallavicini MG, Beisker W, Waldman F. Cell cycle analysis using flow cytometry. International Journal of Radiation Biology and Related Studies in Physics, Chemistry, and Medicine. 1986;**49**(2):237-255

[56] Berridge MV, Herst PM, Tan AS. Tetrazolium dyes as tools in cell biology: New insights into their cellular reduction. Biotechnology Annual Review. 2005:127-150

[57] Namura S, Zhu J, Fink K, Endres M, Srinivasan A, Tomaselli KJ, et al. Activation and cleavage of Caspase-3 in apoptosis induced by experimental cerebral ischemia. The Journal of Neuroscience. 1998;**18**(10):3659-3668

[58] Häcker H, Fürmann C, Wagner H, Häcker G. Caspase-9/−3 activation and apoptosis are induced in mouse macrophages upon ingestion and digestion of *Escherichia coli* bacteria. Journal of Immunology. 2002;**169**(6):3172-3179

[59] Sharifia AM, Eslamia H, Larijanib B, Davoodi J. Involvement of caspase-8, -9,

and -3 in high glucose-induced apoptosis in PC12. Neuroscience Letters. 2009;**459**:47-51

[60] Chen HC. Boyden chamber assay. Methods in Molecular Biology. 2005;**294**:15-22

[61] Millar AJ, Short SR, Hiratsuka K, Chua N-H, Kay SA. Firefly luciferase as a reporter of regulated gene expression in higher plants. Plant Molecular Biology Reporter. 1992;**10**(4):324-337

[62] Himes SR, Shannon MF. Assays for transcriptional activity based on the luciferase reporter gene. In: Tymms MJ, editor. Transcription Factor Protocols. Methods in Molecular Biology. Humana Press; 2000. p. 130

[63] Dien BJ, Gelebart P, Anand M, Zak Z, Hegazy SA, Amin HM, et al. IL-21 contributes to JAK3/STAT3 activation and promotes cell growth in ALK-positive anaplastic large cell lymphoma. The American Journal of Pathology. 2009;**175**(2):825-834

[64] Berglund LJ, Avery DT, Ma CS, Moens L, Deenick EK, Bustamante J, et al. IL-21 signaling via STAT3 primes human naïve B cells to respond to IL-2 to enhance their differentiation into plasma blasts. Blood. 2013;**122**:3940-3950. DOI: 10.1182/blood-2013-06-506865

[65] Pulliam SR, Pellom ST Jr, Shanker A, Adunyah SE. Butyrate regulates the expression of inflammatory and chemotactic cytokines in human acute leukemic cells during apoptosis. Cytokine. 2016;**84**:74-87

[66] Booker BE, Clark RS, Pellom ST, Adunyah SE. Interleukin-34 induces monocytic-like differentiation in leukemia cell lines. International Journal of Biochemistry and Molecular Biology. 2015;**6**(1):1-16

[67] Wang L-N, Cui Y-X, Ruge F, Hang WG. Interleukin-21and its receptor play a role in proliferation, migration and invasion of breast cancer cells. Cancer Genomics & Proteomics. 2015;**12**:22-222

[68] Abdalkareem EA, Ong CY, Lim BH, Khoo BY. Neutralizing FGF4 protein in conditioned medium of IL-21-slilenced HCT116 cells restores the migratory activity of the colorectal cancer cells. Cytotechnology. 2018;**70**(5):1363-1374

[69] Spolski R, Leonard WJ. Interleukin-21: A double edged sword with therapeutic potential. Nature Reviews. Drug Discovery. 2014;**13**(5):379-395

[70] Di Carlo E, Comes A, Orengo AM, Rosso O, Meazza R, Musiani P, et al. Mice IL-21 induces tumor rejection by specific CTL and IFN-γ-dependent CXC chemokines in syngeneic. Journal of Immunology. 2004;**172**(3):1540-1547. DOI: 10.4049/jimmunol.172.3.1540

[71] Avery DT, Deemick EK, Ma CS, Suryani S, Simpson N, Chew GY, et al. B cell-intrinsic signaling through IL-21 receptor for establishing long-lived antibody responses in humans. The Journal of Experimental Medicine. 2010;**207**(1):156-171

[72] Agrawal A, Chen J, Su H, Osann K, Agarawal SG. Increased IL-21 secretion by aged CD4+ cells is associated with prolonged STAT4 activation and CMV seropositicity. Aging (Albany). 2012;**4**(9):648-659

[73] Vallarino AV, Kanno Y, Ferdrinand JR, O'Shea JJ. Mechanisms of Jak/STAT signaling in immune diseases. The Journal of Immunology. 2015;**194**(1):21-27

[74] Yuan M-J, Wang T. Advances of the Interleukin-21 signaling pathway in immunity and angiogenesis. Biomed Reports. 2016;**5**:3-6

[75] Wan C-K, Andraski AB, Solski R, Li P, Kazemian M, Oh J, et al. Opposing

roles of STAT1 and STAT3 in IL-21 function in CD+ T cells. Proceedings of the National Academy of Sciences USA. 2015;**112**(30):9394-9399

[76] Glienke W, Hausmann E, Bergmann L. Down regulation of STAT3 signaling induces apoptosis but also promotes anti-apoptotic gene expression in human pancreatic cell line. Tumor Biology. 2011;**32**(3):493-500

[77] Holm TL, Tornehave D, Sondergaard H, Kvist PH, Sondergaad BC, Hermit MB, et al. Evaluating IL-21 as potential therapeutic target in Crohn's disease. Gastroenterology Research and Practice. 2018;**218**:22. Article ID 5962624

[78] Ngai H, Tian G, Courtney AN, Ravari SB, Guo L, Liu B, et al. IL-21 selectively protects CD62L + NKTcells, and enhances their effector functions for adoptive immunotherapy. The Journal of Immunology. 2018;**201**(7):2141-2153. DOI: 10.4049/jimmunol.1800429

[79] Seo H, Kim B-S, Bae EA, Min BS, Han YD, Shin SJ, et al. IL-21 therapy combined with FD-1 and Tim-3 blockage provides enhanced NK cell antitumor activity against MHC class-I-deficient tumors. Cancer Immunology Research. 2018;**6**(6):685-695. DOI: 10.1158/2326-6066.CIR-17-0708

[80] Solaymani-Mohammadi S, Berzofsky JA. Interleukin-21 collaborates with interferon-γ for the optimal expression of interferon-stimulated genes and enhances protection against enteric microbial infection. PLoS Pathogens. 2019:1-27. DOI: 10.1371/journal.ppat.1007614

[81] Lee IJ, Wu CJ, Wu PY, Tao MH. Radiation improved IL-21 cancer immunotherapy. The Journal of Immunology. 2017;**198**(1 supplement):204-221

[82] Croce M, Rigo V, Sivano F. IL-21: A pleiotropic cytokine with potential applications in oncology. Journal of Immunology Research. 2015. DOI: 10.1155/2015/696578

Chapter 9

Cytokines' Involvement in Periodontal Changes

Petra Surlin, Liliana Foia, Sorina Solomon, Dora Maria Popescu, Dorin Nicolae Gheorghe, Adrian Camen, Maria Alexandra Martu, Anne Marie Rauten, Madalina Olteanu, Allma Pitru, Vasilica Toma, Simona Popa, Mihail Virgil Boldeanu, Silvia Martu and Ion Rogoveanu

Abstract

The bacterial challenge on the periodontal tissues triggers an inflammatory reaction, driven by pro-inflammatory cytokines, that eventually leads to the periodontal structures' damage. The pathogenic mechanisms of this inflammatory reaction are complex and are influenced by the type of host-immune response and certain local and systemic factors. These factors can influence periodontal inflammation, through the action of the various pro-inflammatory cytokines. Periodontal disease and certain systemic conditions can have a mutual association, as the pathogenic mechanisms of these diseases can involve similar molecular and cellular elements. The concept of 'periodontal medicine' comprises these pathogenic connections, focusing on the key role that periodontal health has on the general homeostasis and well-being.

Keywords: periodontal disease, cytokines, inflammation, systemic conditions, associations

1. Introduction

Periodontal disease is defined as an immune, inflammatory disease with a triggering bacterial factor [1]. Its initial phase is characterised by damage to soft gingival tissue (inflammation of gums, gingivitis, plaque accumulation), followed by degradation of the periodontium (chronic inflammation, colonisation by periodontal pathogenic anaerobes) and the surrounding connective tissue matrix of teeth, along with alveolar bone loss, local citrullination, bone resorption and rapid loss of teeth, in the destructive phase [2].

It is considered a multifactorial disease with progressive evolution and pathophysiological mechanisms that involve the association of environmental factors (smoking, stress), microbial factors (several bacterial species, for example, *Porphyromonas gingivalis*, *Treponema denticola*, *Tannerella forsythia*, *Aggregatibacter actinomycetemcimtans*), and genetic polymorphism (human leukocyte antigen-antigen D, HLA-DR, HLA-DRB1; IL-1, IL-6 and IL-10 gene polymorphism) with the inflammatory and immune response (both innate and adaptive immunity) of the host (**Table 1**) [3, 4]. These mechanisms will lead to defective host defences, which will contribute to changes in tissue homeostasis, inflammation and bone loss, the main features of periodontal disease (**Figure 1**) [2].

General function	Type	Specific function	Periodontal implication
Pro-inflammatory	IL-1β	Increases production of other pro-inflammatory mediators (PGE2, IL-6); stimulates neutrophil activity	Key role in the pathogenic processes; enhances alveolar bone-resorption; fuels the inflammatory reaction
	IL-1α	'Alarmin' for tissue damage and immune system; interacts with TNF-α; induces protease synthesis	Enhances alveolar bone-resorption by signalling the presence of bacterial antigens and enhancing the inflammatory response
	IL-6	Regulates cell growth and differentiation: osteoblasts, B and T lymphocytes	Impairs osteoblast growth and function; increases osteoclast formation from monocytes
	IL-18	Increases neutrophil activity and interferon gamma production	Influences lymphocyte Th1/Th2 differentiation; interacts with IL-1β
	IL-33	Activates Th2 and mast cells; stimulates production of IL-5, IL-13 by Th2 cells	Induction of RANKL; induces periodontal damage by stimulation of Th2 cells
	TNF-α	Major regulator of immune cells' activity; involved in the acute phase reaction	Stimulates damage (by osteoclasts) and prevents repair of periodontal tissues (by fibroblast death); starts IL-1β, PGE_2 synthesis
	PGE_2	Increases production of other pro-inflammatory mediators (MMPs); induces fever	Contributes to bone resorption by increasing osteoclast activity
	MMPs (-1, -8, -9)	Enzymatic degradation of collagen (and other extra-cellular matrix proteins). Stimulated by IL-1β and TNF-α. Influences immune cell migration and adhesion	Periodontal damage by impairment of collagen type I production and degradation of structural collagen; causes activation of osteoclasts and damage to connective tissue
Anti-inflammatory	IL-1Ra	Inhibits IL-1β, IL-1α activity by preventing cellular signalling	Limits alveolar bone-resorption; regulates response to antigens (LPS)
	IL-10	Decreases cytokine production by immune cells; reduces inflammatory response	Down-regulates periodontal inflammation by reducing cytokine synthesis in immune cells
	IL-4	Stimulates tissue repair and regulates immunity; regulates differentiation of Th2 cells	Decreases production of Th2 cells, with important implications to periodontal damage
	TGF-β	Enhances epithelial regeneration/repair	Stimulates gingival fibroblast activity

Table 1.
Mediators' role in general and periodontal inflammatory reactions [3].

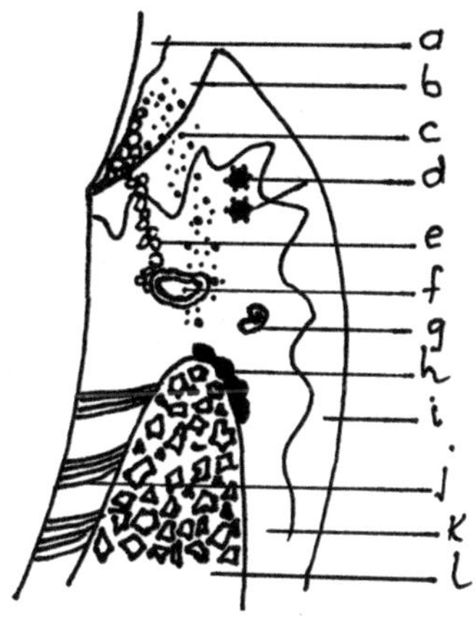

Figure 1.
The periodontium and its inflammatory reaction in periodontal disease: a—bacterial plaque deposits; b—gingival sulcus; c—inflammatory mediators (interleukins—ILs, matrix-metalloproteinases—MMPs, tissue inhibitors of MMP—TIMPs); d—lymphocytes type B and T; e—polymorphonuclear cells; f—blood vessel; g—macrophage cells; h—osteoclast cells; i—gingival epithelium; j—periodontal ligament; k—gingival connective tissue; and l—alveolar bone.

Pathophysiological processes are explained by the participation of a wide range of locally released soluble factors such as pro-inflammatory cytokines, prostaglandin E2 and reactive oxygen species, inflammatory mediators that can be highlighted by various methods within gingival tissues and within the gingival crevicular fluid (GCF) [5]. The association between elevated TNF-α and IL-6 concentrations and disease activity was also highlighted [6]. It has been shown that IL-1β and interferon-gamma (IFN-γ) have elevated concentrations in active periodontal lesions [7]. IL-12, a key-acting cytokine mediating Th1 differentiation, also implicated in cell-mediated immunity, has been observed to stimulate pro-inflammatory proteins involved in bone resorption [8].

A deficiency in the synthesis and release of anti-inflammatory cytokines can also occur, like type Th2 cytokines (IL-4, IL-5, IL-13), IL-10 and transforming growth factor beta 1 (TGF-β1), which confirms the existence of an imbalance between pro- and anti-inflammatory mechanisms at periodontal level [9]. Along with TNF-α, IL-1β and IL-6, the release of the cytokines IL-8, IL-11, and IL-17 has also been revealed within GCF, which has potent osteoclastogenesis stimulating effects and also reduces osteoprotegerin synthesis in osteoblasts and stromal cells [10].

Periodontal disease can be influenced by certain local factors, such as orthodontic therapy and the existence of coronal or prosthodontic restorations, and by certain systemic diseases, this may interfere with the periodontal inflammatory reaction, by means of pro-inflammatory cytokines. These diseases include diabetes mellitus (type 1 and 2), cardiovascular diseases, rheumatoid arthritis and hepatic and renal conditions. The interactions between periodontal tissues and their pathology and these systemic diseases have been reunited under the concept of 'periodontal medicine' [11]. Considering this concept, the chapter aims to exhibit the major implications of cytokines into the pathogenic mechanisms of periodontal disease and its local and systemic influencing factors.

2. Methodology

The design of the chapter has been created in order to reflect the role of cytokines in periodontal pathogenic processes, as both local and systemic risk factors for periodontal disease have been taken into consideration. The information is divided between the influence of local elements—orthodontic and restorative treatments, and systemic ones—diabetes mellitus type 1 and 2, cardiovascular, rheumatic, hepatic and renal diseases. Relevant scientific information has been sourced from the existing scientific literature and structured so as to pursue the established purpose of the chapter.

3. Local influences on the periodontal inflammatory reaction

3.1 Orthodontic influences

The orthodontic dental movement is the consequence of the application of controlled mechanical forces on the teeth [12], the periodontal ligament (PDL) being the one that mediates the task to which the teeth and the alveolar bone are subjected and creates the conditions for the cells to participate in bone remodelling [13]. Orthodontic forces distort the PDL matrix, causing the cellular form and cytoskeleton configuration to change, with neuropeptides release from the afferent nerve terminals [14], while the orthodontic forces can induce a biomolecular-level release of growth factors, prostaglandins and pro-inflammatory cytokines as IL-1, IL-6, IL-8 and TNF-α, that affect alveolar bone remodelling [15]. At present, the molecular mechanisms underlying bone formation, induced by stretching forces, are not fully understood [16]. The initial phase of orthodontic dental movement always involves an aseptic acute inflammatory reaction [17], which lasts for 1–2 days and is predominantly exudative, followed by a chronic, mainly proliferative process [18] and increased release of cytokines [19], such as TNF-α, TNF-β, IL-1, PDGF, INF-γ and RANKL [15, 20].

Inflammatory cytokines are involved in all phases of inflammation during orthodontic treatment [21]; pro-inflammatory and anti-inflammatory cytokines act synergistically or antagonistically on each other [22]:

- *IL_1 family*: all 3 ligands of the IL-1 family (IL-1α, IL-1β, and IL-1RA) are involved in bone metabolism and orthodontic dental movement [22], as well as IL-1 gene polymorphisms [23]. IL-1β along with TNF-α is a key pro-inflammatory cytokine in acute-phase inflammation [24]. IL-1β gathers leukocytes and activates fibroblasts, endothelial cells, osteoclasts and osteoblasts in order to stimulate bone resorption and to stop bone formation [25]. IL-1β is also a PGE inducer and, along with mechanical stress, synergistically regulates the formation of PGs in periodontal cells [26]. The levels of IL-1β and PGE2 are higher in the tension zones compared to compression zones, which supports the hypothesis that during the initial stage of orthodontic treatment, this cytokine would originate from osteoclasts in response to mechanical stress [27].

- *IL-1RA* acts by limiting the inflammatory conditions [28] mediated by IL-1 and bone resorption [22], and therefore positive correlations exist between decreasing IL-1RA levels in GCF and faster bone resorption during orthodontic dental movement and consequently a higher dental movement speed [29].

  The tooth movement speed is influenced by stress and by levels of IL-1RA, IL-1β and IL-1 gene polymorphisms from GCF. These factors provide a better predictive model for the efficiency of dental movement: activity index [AI = experimental (IL-1β/IL-1RA)/control (IL-1β/IL-1RA)], IL-1RA concentration in GCF and IL-1β genotype [29].

- *IL_6*: IL-6 has a stimulating effect on bone remodelling and osteoclast formation [30] and also in the inflammation associated with orthodontic dental movement [31].

- *IL_8*: IL-8 has a role in the neutrophils' recruitment and activation in the presence of inflammation [32], in improving RANKL expression and consequently in increasing the osteoclast production and their activation [33]. Immediately after the application of mechanical forces, IL-8 has an increased level in both pressure and tension zones [34], but later on, the stimulation of IL-8 secretion only continues in the tension zones; this differential regulation probably plays a major role in the initial stage of bone remodelling [34].

- *TNF-α*: TNF-α binds with macrophage colony-stimulating factors to induce osteoclast differentiation [35].

- *IFN-γ*: IFN-γ increases during late stages of orthodontic dental movement [35], controlling massive osteoclastogenesis [36] related to the increased volume of trabecular bone [37].

The levels of cytokines in the GCF vary with the type and intensity of the applied force, speed of tooth movement, the age of the orthodontic device owner [24] and growth [38]. Equivalent force systems during orthodontic dental movement induce an individualised production of different cytokines [39]. Light continuous forces tend to maintain relatively high levels of IL-1β [29], needed for the continuation of periodontal remodelling, longer periods allowing a reduced frequency of reactivations [38]. A strong force may increase the risk of root resorption and hyalinization of PDL and may also modify the cytokines' level, causing unwanted tissue reactions and the need for multiple reactivations [40]. The increase of IL-1β's level in GCF, as a result of the increased applied force, was also associated with intense pain during orthodontic dental movement [38], probably due to the correlation between IL-1β and substance P [41]. Most studies did not detect differences between the levels of cytokines in the tension and pressure zones [29, 34], probably due to the continuous circulation of GCF in the periodontal ligament. Consequently, it can be concluded that GCF cytokine levels cannot be specific indicators of periodontal remodelling in tension and pressure zones [42].

External apical root resorption (EARR), secondary to orthodontic dental movement, is a frequent clinical complication, and treatment variables, environment factors and/or inter-individual genetic variations may give susceptibility or resistance to its occurrence [43]. Among pro-resorptive cytokines, IL-6 promotes osteoclastic function [44] and amplifies the production of fibroblasts from gingival fluid and periodontium [45], having increased levels in patients with severe dental root resorption [46]. IL-7 acts indirectly on osteoclastogenesis by induction of TNF-α [47]. The increase of IL-6, IL-7 and TNF-α level might be a sign of a continuous remodelling of the periodontal tissues during the lag phase of tooth movement and a mechanism of cellular prohibition [48].

Regarding anti-resorptive cytokines, the levels of IL-4 and IFN-γ increase closely following IL-1β's elevation. This cytokine's expression type may result as an active combined remodelling of periodontal tissue, during the first stages of dental movement and also of cellular prohibition mechanisms, preventing activation and additional differentiation of osteoclastic cells [48] and suppressing osteoclastogenesis, unlike T cells of RANKL that induce osteoclastogenesis [49]. Granulocyte-macrophage colony-stimulating factor (GM-CSF) is also an anti-resorptive cytokine that prevents bone resorption together with IL-4, IL-10, IL-13, IL-18 and IFN-γ [50].

3.2 Restorative influences

The interplay between periodontology and restorative dentistry exists on many levels and functions, in both directions [51]. For example, interproximal restorations, fixed prosthesis and artificial crowns can be involved in the occurrence and progression of gingival inflammation and periodontal destruction [51]. The presence of dental restorations or other appliance, near or below the gingival margin, which reach deep into the gingival sulcus or within the junctional epithelium, may induce localised inflammation that could lead to future periodontal complications [51].

The most reliable mechanism by which the subgingival margins of restorations lead to gingival inflammation and loss of attachment is the increase of plaque accumulation. In these conditions, the rate flow of gingival fluid will increase, because the gingival fluid protects oral tissues (including the junctional epithelium) against the bacterial invasion, acting as a defensive barrier, within the gingival sulcus [52]. IL-1, prostaglandin E2 (PGE_2) and the elements that affect collagen and bone (matrix metalloproteinases (MMPs)) that are detected in gingival fluid can be used as possible indicators for the diagnosis of periodontal disease and its progression [2]. The placement of restorative posts can induce trauma to periodontal structures, mostly in the case of metallic posts and less for the fibre ones, leading to periodontal inflammation [53]. IL-1 family cytokines regulate the activity of other pro-inflammatory cytokines, such as PGE2 and MMPs, their gingival fluid levels being proportionate with the degree of bone loss [10].

The placement of the gingival margins of restorations depends on the position of the decayed portion of the tooth and the extent of caries. The materials used in restorations must not be injurious to periodontal tissues, as they should have highly polished surfaces, the periodontium reacting to the roughness of the material and its accumulation of plaque by inflammation. However, it has been shown that the composition of restorative materials could also initiate periodontal changes, by monocyte activation and changes of the gingival fluid levels of cytokines [54]. It has been observed that periodontal inflammation can occur in areas of the gingival tissue which are adjacent to ceramic, composite or amalgam coronal restorations, even if there are no signs of bacterial plaque accumulation. This may be explained by the stimulation of periodontal neurogenic inflammatory reactions that is caused by these restorative materials. However, these findings are controversial and require further research, as it has been reported that only class V composite fillings can trigger such forms of gingival inflammation, in the areas of their vicinity [55].

Ceramic materials have the lowest plaque retentive capacity, but inner materials of porcelain-fused-to-metal (PFM) crowns can impact periodontal tissues [56]. Certain non-noble metals could have a negative impact on periodontal tissue (Ni-Cr alloy, Co-Cr alloy). It seems that the Au-Pt alloy is least harmful for the periodontium. The spaces between the margins of the restoration, as well as their contour are other contributing factors to gingival inflammation and periodontal destruction, referred to as iatrogenic factors. Restoration margins have been reported as key factors for periodontal health [57]. Another feature referring to periodontal damage is the dental impression that needs to offer a clear image and perspective of the prepared teeth, the neighbouring teeth and the associated gingival tissue. Impressions of tooth preparations with an elastic material that extends subgingivally could damage soft tissues. In order to prevent this, the retraction cord is frequently used during impression taking [58]. Successful restorative and prosthetic treatments require a healthy periodontium, as a start point of therapeutic protocols [51].

4. Systemic pathologic influences on the periodontal inflammatory reaction

4.1 Type 2 diabetes

The relationship between type 2 diabetes mellitus (T2DM) and chronic periodontitis is bidirectional [59]. The deleterious effect of chronic hyperglycemia on the occurrence and progression of periodontitis has been observed in patients suffering from both diseases [59]. Among the factors involved in the pathogenesis of periodontitis in patients with T2DM are the advanced glycation end products, chronic inflammation, altered secretion profile of cytokines, decrease of the immune response to infection and also the exacerbation of bacterial flora at periodontal level [59].

Recent evidence has suggested the role of periodontitis in the pathogenesis of insulin resistance, the adverse effects of periodontitis on glycemic control, and the contribution to the occurrence and progression of chronic diabetic complications [60]. The link between periodontitis and T2DM is potentially explained by the increase of pro-inflammatory cytokines, such as IL-1β, IL-8, IL-6, TNF-α, IL-12, and leptin and decrease of anti-inflammatory cytokines such as IL-4, IL-11, adiponectin and fibroblast growth factor-21 [60, 61].

The assessment of cytokine levels within the gingival crevicular fluid indicated that subjects with T2DM and chronic periodontal disease have significantly higher levels of IL-1β, IL-6 and TNF-α than healthy participants with similar periodontal conditions [59]. The levels of IL-8 within the gingival crevicular fluid and plasma have been evaluated with contradictory results in the context of T2DM and chronic periodontitis, multiple studies reporting that the IL-8 level is higher in patients with T2DM and chronic periodontitis compared to non-diabetic subjects with or without chronic periodontitis [60, 62].

The systemic inflammatory response depends on the ratios of pro-inflammatory to anti-inflammatory cytokines. The patients with T2DM and chronic periodontitis tend to have higher ratios of pro-inflammatory to anti-inflammatory cytokines compared to periodontally and systemically healthy controls [63]. Thus, TNF-α/IL-4, IL-1β/IL-4, IL-23/IL-4, IL-6/IL-4, TNF-α/IL-5, IL-17/IL-5 and IL-6/IL-5 ratios were higher in T2DM patients with chronic periodontitis than in healthy subjects [64].

The literature regarding the systemic plasma cytokine profile's characteristics in patients with T2DM and chronic periodontitis is not consistent. Some reports indicated lower levels of cytokines, including IL-4, IL-5, IL-6, TNF-α, IL-1β, IL-17, IL-13, IFN-γ, IL-2, IL-23 and IL-12 and higher levels of IL-8 in serum samples of T2DM patients with chronic periodontitis than for the control group [60, 63]. There are other studies showing that the systemic levels of some cytokines, such as IL-4, IL-8, IL-6, IL-10 and TNF-α, did not differ between diabetic and non-diabetic subjects with periodontitis [62, 64].

The improvement of glycemic control after periodontal therapy may be explained by the fact that periodontal treatment relives the periodontal inflammatory reaction and further decreases systemic pro-inflammatory cytokines' involvement in the pathogenesis of insulin resistance [65]. Reduction of the pro-inflammatory cytokines such as IL-1β and IL-6 in gingival crevicular fluid and improvement of glycemic control after periodontal therapy was reported in T2DM patients [66]. Furthermore, a significant decrease in serum levels of IL-6 and TNF-α and also a reduction of HbA1c in patients with T2DM were reported after periodontal therapy, indicating the impact of periodontal intervention on periodontal and systemic inflammation related to insulin resistance [65, 66]. Recent findings show that reducing periodontal inflammation by periodontal therapy may contribute to an increase of systemic anti-inflammatory cytokines, such as adiponectin and fibroblast growth factor-21 levels, and to a decrease of leptin levels, thus improving insulin sensitivity [65].

As high levels of pro-inflammatory cytokines have been reported at periodontal level, associated with increased risk of destructive effects within the periodontal tissues [67], it was shown that diseases characterised by insulin resistance like T2DM are also associated with the increase of cytokines [68, 69]. Chronic periodontitis may influence systemic cytokines in T2DM. The literature documents the role of cytokines TNF-α, IL-1β, IL-4, IL-6 and IL-10 in chronic periodontitis and T2DM [70].

Pathogenic links between periodontal disease and diabetes involve elevations in IL-1β [71], TNF-α, IL-6, RANKL-b, oxidative stress and Toll-like receptor (TLR) expression. It has been demonstrated that prolific circulatory mediators have higher levels in the association of diabetes with a form of periodontal disease, especially TNF-α, C-reactive protein (CRP) and mediators of oxidative stress, which in turn can affect the control of diabetes. Moreover, complete and correct periodontal treatment can improve serum levels of CRP and TNF-α in patients affected by diabetes [65, 72].

The release of inflammatory mediators that can be detected in GCF stimulates the secretion of metalloproteinases, initiates bone resorption processes and plays an important role in the evolution and prognosis of periodontal disease. Inflammatory cytokine and chemokine levels in GCF decreased after initial periodontal therapy [73]. In periodontitis, concentrations of pro-inflammatory cytokines TNF-α and IL-8 increased not only in periodontal tissues [74] but also in serum samples [62].

4.2 Type 1 diabetes

The existence of a bidirectional relationship between periodontal disease and type one diabetes (T1D) has long been considered [75]. Periodontal breakdown is a pivotal aggravating factor for the health status in subjects with T1D, mainly because this preserves a chronic systemic inflammatory condition, contributing therefore to diabetic complications [75]. In fact, periodontitis has been considered to be the sixth complication of diabetic disease, diabetic metabolic impairment being in turn able to point toward a poor periodontal health status. Being not only a metabolic misbalance, but also tightly associated to an important dysfunction of the immune system, several facets of the systemic immune response, such as antigen challenge or polymorphonuclear leukocyte and T-lymphocyte function are altered in diabetic subjects [76].

Hyperglycemia can lead to immune system disorders; hence, the effects mediated by cytokine alterations in patients with T1D might be significant. IL-1β levels in gingival fluid and those of IL-6 in saliva have been correlated with glycosylated haemoglobin [77]. The changes in the cytokine amount on the systemic level are crucial in the pathogenesis of diabetes and can condition the islet cell turnover and apoptosis, with subsequent disease progression toward devastating complications, as macro- and microvascular modifications [78].

Regarding the involvement of cytokines in the immune-inflammatory response in subjects with T1D and periodontal breakdown, the literature highlights the enhancement of IL-1β, IL-6 and prostaglandin E2 in the gingival fluid of these patients, compared to systemically healthy subjects, with comparable periodontal alteration [79, 80]. The results on experimental murine models of induced diabetes include data on a broader range of mediators: IFN (interferon), chemokines such as macrophage inhibiting protein (MIP-2) and monocyte chemo-attractant protein (MCP-1), most likely mediated via TNF-α [81].

The importance of TNF for enhancing the immune response, generated by bacterial plaque challenge, in T1D and T2D, has been experimentally shown for murine models with chronic periodontitis, but without a clear relation between the TNF-α levels found in oral tissues and those found in oral fluids of T2D periodontal subjects [81]. Along with these pro-inflammatory mediators, research has also been carried upon chemokines' performance, growth factors and soluble adhesion molecules, which possess immune-regulatory capacity [79]. The simultaneous action of multiple mediators in evaluating oral immune response in periodontal patients with T1D has also been assessed [82].

Monocytes can have a hyper-inflammatory phenotype in patients with T1D, and these cells are susceptible to the action of lipopolysaccharides (LPS) of the periodontal bacteria, and respond through generation of increased amounts of IL-1β, TNF-α and PGE2, compared to non-diabetic subjects [79]. This inflammatory phenotype of monocytes represents one of the relevant links between periodontal pathogenesis and diabetes mellitus [79].

Research on other cell populations involved in mediating the immune response advocates that T cells accumulation around insulin-sensitive cells is important in metabolic changes correlated to diabetes, via their capacity to moderate macrophage activity [83]. There are a number of distinct subclasses of T cells, with remarkable plasticity, their role being strongly correlated to the local cytokine amount. However, although important in maintaining the Th1/Th2 cell balance in the pathogenesis and progression of periodontitis, data on the underlying mechanisms and their role in homeostasis of periodontal status in patients with diabetes are still to be elucidated [76].

Gathering the whole data related to the modulation of locally expressed mediators, the T1D-periodontitis association would be of particular relevance in the curative management plan, through conduction of long-term studies. Further research is necessary to better understand the cytokine expression in periodontal disease and type 1 diabetes.

4.3 Cardiovascular diseases

Cardiovascular disease (CVD) constitutes an extensive cluster of conditions that deter the physiological function of the heart and/or blood vessels and includes (a) coronary heart disease—angina/myocardial infarction; (b) ischaemic cerebrovascular disease—transient ischaemic attack—TIA and (c) peripheral vascular disease. In both cardiovascular disease and periodontal disease, a systemic inflammatory overload is present in the organism, which can be exacerbated by factors such as smoking, obesity or diabetes mellitus that eventually leads to an altered dishomeostatic status [84]. Due to the ulceration and inflammation of subgingival epithelial layer, oral pathogens and their by-products can access the blood stream, increasing the risk and aggravating the evolution of any pre-existing heart disease [85].

Atherosclerosis is commenced by deterioration of the endothelial tissue of vessels. Following impairment, endothelial pro-inflammatory signals drive not only the expression of adhesion molecules such as E-selectin, ICAM-1, P-selectin and VCAM-1, but also of IL-8 and thrombin that act as chemoattractants and determine an upsurge in the aggregation of platelets and migration of leukocytes [86]. Furthermore, these pro-inflammatory signals trigger the proliferation of smooth muscle cells and the apoptosis of endothelial cells. In this context of complex interactions, leukocytes migrate to the injured site and release additional pro-inflammatory cytokines (IL-1α and β, IL-6, IL-17, IL-22 and TNF-α), reactive oxygen species, as well as proteinases that break down the extracellular matrix of the endothelium [87].

High concentrations of low-density lipoproteins (LDL) in plasma accumulate in the aortic wall beneath the intima layer of the endothelium and are oxidated into oxLDL. Increased blood pressure activates endothelial cells which advocate the activation of adhesion molecules that favour the migration of monocytes. In the aortic wall, these cells transform into macrophages that absorb the oxidated LDL and develop into foam cells filled with lipids. This circumstance further promotes the dispersion of pro-inflammatory molecules which stimulate the invasion and activation of supplementary inflammatory cells and facilitate their confinement in the plaque, precipitating subsequent build-up of inflammation factors [88].

Periodontal microorganisms can alter the mentioned mechanisms by virtue of a direct interaction (for example by the invasion of endothelial, smooth muscle cells, leukocytes and platelets), but also an indirect interplay by stimulating the release of paracrine factors that eventually affect normal cellular function [89]. In addition to direct invasion, microorganisms can release products into the circulation and induce pro-atherogenic responses in endothelial cells [90]. Research has highlighted that vesicles pertaining to the outer membrane such as gingipains from *P. gingivalis* and free soluble components deriving from *A. actinomycetemcomitans* cause irritation to endothelial cells and promote inflammation [91]. Effusion of pro-inflammatory cytokines (IL-1, IL-6 and TNF-α) and other chemokines from the affected periodontal tissues leads to the generation of acute phase proteins (C reactive protein, fibrinogen, amyloid A, etc.) by the liver [92]. As a result of this injury, the activation of the adaptive immune system occurs as a typical response to chronic bacteraemia. Pathogen-associated molecular patterns (PAMPs) of periodontal pathogens determine the production of antibodies and elicit a cross-response between endothelial cells and the altered LDL to exacerbate the migration of lipids into cells inside the blood vessel walls [93]. Antigen-presenting cells, for example, dendritic cells and effector T lymphocytes, bear a substantial part in the generation of proatherogenic cytokines, for example, IL-20, IL-18 and IFN-γ, and are consequently also relevant in the progression of atherosclerotic plaques. A noteworthy role among the proatherogenic cytokines has been assigned to IL-12 due to the fact that when this certain molecule was absent early lesion development was inhibited. However, late progression was unaffected [94].

Considering the recent epidemiological, experimental and clinical evidence presented in the literature on this subject matter [86], the issue of an interrelationship between periodontal disease and cardiovascular disease is clearly supported. In spite this, the exact nature of this relationship, whether it is a direct or causal one, remains to be edified through interventional and longitudinal studies.

4.4 Rheumatoid arthritis

Rheumatoid arthritis (RA) is considered to be an autoimmune, progressive, inflammatory and chronic disorder, in which the human immune system reacts in an erroneous manner to the articular structures. This response is characterised by an inflammatory immune cell infiltrate (activation of innate immunity, followed by the emergence of adaptive immune responses) in the early phase, followed by a destructive phase characterised by degenerative phenomena, such as denaturation of normal synovial structures by hyperproliferation, reactive fibro-vascular proliferation, bone destruction and cartilage degradation. This explains the occurrence of swelling and pain within and around joints [95].

Over the past two decades, RA has been described as a disease model in which various pathophysiological mechanisms have been studied to better explain the inflammatory process, but also the modality of involvement of the human immune system, through two types of specific responses [95]. These research projects culminated with the introduction into patients' treatment regimens of the chimeric anti-TNF-α monoclonal antibodies (infliximab, 75% human and 25% mouse peptide sequences) [96], a recombinant human TNF receptor (p75)–Fc fusion protein (etanercept) and biological therapy with fully human anti-TNF-α monoclonal antibodies [97].

Major cytokines, such as TNF-α and interleukins (IL-17 and IL-1β), function by promoting inflammatory responses, causing inflammation of the synovium and inducing cartilage degradation. Other soluble mediators, such as cytokines released by the Th2 lymphocyte subpopulation (the most frequently studied and for which statistically significant results were obtained, would be IL-4, IL-10, IL-13) are mainly anti-inflammatory molecules [95, 98]. In addition, IL-13 could have potential clinical importance because it can suppress both secretion and actions of IL-17 [98].

Over the past two decades, the existence of links or associations between RA and periodontitis has been investigated [99]. RA could be a triggering risk factor for periodontitis as there is a high incidence of RA in patients with periodontal disease, and RA patients have more chances to experience moderate to severe periodontal changes, as compared to healthy subjects. It has also been found that calcifications and soft tissue injuries have the same characteristics, leading to the conclusion that the destructive inflammatory chronic lesions are similar [100].

New theories have highlighted that periodontal disease is a risk factor for RA, with several similarities existing between RA and periodontitis [101–103]:

- both diseases are multifactorial, chronic immune-inflammatory diseases with progressive evolution;

- pathophysiological mechanisms involve the association of environmental factors, microbial factors, genetic susceptibility (it was found that the HLA-DRB1 subtype is associated with both fast progressive periodontitis as well as RA);

- locally, the cellular composition of the inflammatory reactions is similar, it includes both the T and B subtypes of lymphocyte populations, but also the cells involved in the activation of lymphocytes such as dendritic cells;

- for both diseases, it was found that the changes occurring in early phases, as well as those in destructive phases are characterised by degenerative phenomena and periodontal degradation, but also the chronic systemic inflammation is determined and mediated by the interruption of balance between the pro-inflammatory and the anti-inflammatory cytokines; and

- change in connective tissue and bone homeostasis is irreversible, consisting in the deterioration of collagen-rich structures, a process in which they intensively participate and have collagenolytic effects as matrix metalloproteinases and other enzymes (elastase, bacterial cysteine proteases, enzymes associated with neutrophils).

4.5 Liver diseases

One of the most damaging hepatic diseases is chronic hepatitis C (CHC), occurring after the infection with the hepatitis C virus (HCV), which replicates within hepatic and peripheral blood cells [104]. It is considered that worldwide, more than 200 million persons are affected by this disease, making it a major public health concern [105]. As the hepatic inflammation progresses and becomes chronic, healthy liver tissue is replaced by fibrotic tissues, which is unable to perform normal hepatic functions, leading to cirrhosis [106]. During chronic hepatic inflammation important cytokine profile changes can be observed in affected patients, with increased levels of pro-inflammatory markers, such as TNF-α, being found in their serum samples [107].

Cytokine levels, measured in either gingival fluid or serum samples, can be used to assess the progression and severity of both periodontal and hepatic disease [108]. In periodontal disease, the secretion rate of gingival fluid increases, as well as its content in pro-inflammatory cytokines [109]. The levels of these cytokines can be determined, as an indirect indicator of the disease's severity and progression rate [110]. Being easy to sample, the gingival fluid has allowed extensive research on its pro-inflammatory cytokine content [111]. Similarly, the evolution of chronic hepatitis C can be monitored by the assessment of some cytokine's levels (such as IL-18 and IL-33) in the serum samples of affected patients [112, 113]. Moreover, the same pro-inflammatory cytokines, such as IL-1, Il-6 and interferon-gamma, can express elevated levels during chronic hepatic inflammation (as in chronic hepatitis C), as well as during chronic periodontal inflammation (as in chronic periodontal disease) [114]. The similar profiles of these cytokines suggest that the two inflammatory reactions could be driven by the same pro-inflammatory markers, endorsing extensive research on their possible common pathogenic mechanisms [108].

Another frequently encountered hepatic condition, particularly within developed regions, is non-alcoholic fatty liver disease (NAFLD). Affecting almost 24% of the global population [115], the disease is mainly caused by genetic or behavioural factors, such as a misbalanced diet, rich in lipids and sugars, lack of physical activity or can occur during other systemic conditions like obesity and diabetes mellitus [116]. The accumulation of fat leads to hepatic steatosis, which, in time, can develop life-threatening complications, as liver cirrhosis or hepatic cancer [117]. NAFLD has been associated with metabolic disorders, as insulin resistance [118], as the adipocytes of the fatty tissue will produce increased levels of TNF-α [119]. This cytokine can alter cellular sensitivity to insulin, therefore decreasing normal glucose metabolism. As a result, serum glucose will rise, increasing the risk of diabetes mellitus [118].

TNF-α is also intensively involved in periodontal inflammation. This cytokine has been shown to express increased levels in the gingival fluid of periodontal patients [120] and has similar pro-inflammatory mechanisms to IL-1β [121]. Its main damaging role during periodontal inflammation is the disruption of the normal reparatory function of fibroblasts and the stimulation of bone-resorption osteoclasts [122]. TNF-α may represent a key cytokine that links the pathogenic mechanisms that are common between periodontal disease and insulin resistance, insulin resistance and NAFLD, and consequently between periodontal disease and NAFLD [123].

The relationship between periodontal disease and hepatic conditions could be considered as functioning into two directions. Firstly, periodontal disease can impact the development and outcome of hepatic conditions. The main explanation of the root cause of the association between periodontal disease and hepatic conditions seems to be the bacterial challenge that leads to periodontal inflammation [124]. Pre-existing hepatic conditions, as chronic hepatitis, NAFLD or cirrhosis can be aggravated by the impact of bacterial attack on the periodontal tissues [125]. Important periodontal pathogens, like *Porphyromonas gingivalis* and *Aggregatibacter actinomycetemcomitans* stimulate the synthesis of pro-inflammatory cytokines by periodontal cells [126]. These cytokines, including TNF-α and IL-1 family ones, will be carried by the vascular system and reach the liver. Here, they can have damaging effects on the hepatocytes, if there is a pre-existing hepatic condition, by adding to the distress of the already impaired hepatic tissue [127]. Consequently, the liver functions will be more difficult to perform and the conditions will aggravate, as a result of a bacterial periodontal challenge [128].

Conversely, hepatic conditions can impact the evolution and manifestations of periodontal disease, by means of increased cytokine production [129]. Various liver diseases, as NAFLD, CHC and liver cirrhosis, trigger the increased production of pro-inflammatory cytokines, including TNF-α, IL-1 family, IL-6, which also have a proven active role in the promoting of the periodontal inflammatory reaction [130]. These cytokines, originating from the affected hepatic tissue, enter the blood stream and reach the periodontal tissues. When a periodontal bacterial challenge occurs, they contribute to the exacerbation of the inflammatory reaction, inflicting damage on the periodontal tissues, together with the periodontal-originating pro-inflammatory cytokines. Therefore, an exaggerated inflammatory response is triggered, causing important loss of periodontal structures in patients who also suffer from chronic liver diseases [131, 132].

4.6 Renal diseases

Chronic kidney disease (CKD) is an official public health concern, with 10–12% of the population affected in terms of mortality and morbidity [133, 134]. CKD is characterised by the use of certain markers that indicate the degree of kidney malfunction, notably the glomerular filtration rate (GFR), which indicates the kidney's functioning efficiency. CKD is diagnosed when multiple standard criteria are met, including a GFR lower than 60 mL/min, albumin levels higher than 30 mg/g of creatinine and the existence of morphological kidney changes. When the GFR drops below 15–20 mL/min, it can be considered that end-stage renal disease has occurred [135, 136].

Various harmful stimuli are triggers for inflammation, the physiological protecting mechanism of the body. In CKD, as in several other chronic debilitating disorders, inflammation becomes maladaptive, uncontrolled and persistent. In this group of patients, a majority of the patients with minimum Stage 3 CKD have increased levels of C-reactive protein (CRP) [137], this prevalence being even higher either in final stage CKD or in dialysis patients [138]. For the evaluation of inflammatory state in clinical practice, a series of specific markers are used. One of the most important inflammation indicators is the CRP, which can also be found in its high sensitive form (hs-CRP) in elevated levels in the serum samples of patients with chronic renal failure, along other pro-inflammatory cytokines [139]. Moreover, another pro-inflammatory cytokine, IL-6, could be a reliable indicator of the risk of cardiovascular diseases and mortality in subjects with end stage renal disease (ESRD) [140].

Malnutrition, a severe consequence of CKD, inflicts important changes in most ESRD patients, in terms of anthropometric and serologic aspects, most of which have an irreversible character, even with proper nutritional supplementation. In addition, malnutrition also comprises a chronic inflammatory reaction, driven by pro-inflammatory cytokines (IL-1, IL-6, TNF-α, IFN-γ, etc). This immune response can accelerate the muscular protein catabolism, on the one hand, by elevating the hepatic synthesis of positive acute phase proteins and, on the other hand, by suppressing the production of negative acute phase proteins [141].

ESRD and haemodialysis (HD) itself lead to an inflammatory status, influenced by numerous factors. The main factor of morbidity and mortality in dialysis patients is considered chronic inflammation, a major determinant of 'dialysis syndrome'. Inflammation in dialysed patients is characterised by enhanced production of CRP, TNF-α, IL-6, IL-2 and chemokines, such as IL-8, and it may vary over time and during this process [142].

The bacteria of the subgingival biofilm can reach the systemic blood circulatory system, causing lesions to the arterial endothelium and triggering a series of pathogenic events that can eventually lead to atherosclerosis [85]. This mechanism can explain certain pathogenic connections between local and systemic conditions. These connections may have a bidirectional nature, deriving from the similar local and systemic inflammatory reactions that these conditions manifest [143]. For instance, a renal inflammatory response (glomerulonephritis) can be triggered by various acute and chronic infections. This is supported by the increased prevalence of periodontal pathology in patients with renal diseases, suggesting the high relevance of the periodontal pathology in the onset of renal inflammatory responses [144].

Recently, research has focused on the contribution of inflammation, determined by periodontitis, to the overall inflammatory systemic burden in patients with CKD and dialysis. In this context, plasma levels of certain inflammatory cytokines may be relevant to malnutrition, morbidity and mortality of these patients, as well as to their quality of life [145]. Individuals with ESRD experience reduced quality of life and many of them also associate low oral health [145].

5. Conclusion

Through the intense implication of cytokines, the periodontal inflammatory reaction can exhibit a variety of clinical manifestations, in terms of onset, evolution, treatment and prognosis. The local and systemic factors, which can influence the development of periodontal inflammation, often have molecular and cellular implications, and cytokines become the means through which different pathologies can have mutual impact. The cytokine standpoint on periodontal disease and its connective systemic conditions offers wide and promising perspectives on further developments of more precise methods of diagnosis and more efficient therapeutic protocols.

Conflict of interest

None.

Author details

Petra Surlin[1], Liliana Foia[2], Sorina Solomon[3], Dora Maria Popescu[1], Dorin Nicolae Gheorghe[1*], Adrian Camen[4], Maria Alexandra Martu[5], Anne Marie Rauten[6], Madalina Olteanu[6], Allma Pitru[7], Vasilica Toma[5], Simona Popa[8], Mihail Virgil Boldeanu[9], Silvia Martu[3] and Ion Rogoveanu[10]

1 Department of Periodontology, Faculty of Dental Medicine, University of Medicine and Pharmacy of Craiova, Craiova, Romania

2 Department of Biochemistry, Faculty of Dental Medicine, "Grigore T. Popa" University of Medicine and Pharmacy, Iasi, Romania

3 Department of Periodontology, Faculty of Dental Medicine, "Grigore T. Popa" University of Medicine and Pharmacy, Iasi, Romania

4 Department of Oral Surgery, Faculty of Dental Medicine, University of Medicine and Pharmacy of Craiova, Craiova, Romania

5 Department of Pedodontics, Faculty of Dental Medicine, "Grigore T. Popa" University of Medicine and Pharmacy, Iasi, Romania

6 Department of Orthodontics, Faculty of Dental Medicine, University of Medicine and Pharmacy of Craiova, Craiova, Romania

7 Department of Oral Pathology, Faculty of Dental Medicine, University of Medicine and Pharmacy of Craiova, Craiova, Romania

8 Department of Diabetes, Nutrition and Metabolic diseases, Faculty of Medicine, University of Medicine and Pharmacy of Craiova, Craiova, Romania

9 Department of Immunology, Faculty of Medicine, University of Medicine and Pharmacy of Craiova, Craiova, Romania

10 Department of Gastroenterology, Faculty of Medicine, University of Medicine and Pharmacy of Craiova, Craiova, Romania

*Address all correspondence to: dorinngheorghe@gmail.com

References

[1] Scott DA, Krauss J. Neutrophils in periodontal inflammation. In: Kinane DF, Mombelli A, editors. Periodontal Disease. Vol. 15. Basel: Karger Publishers; 2012. pp. 56-83

[2] Cekici A, Kantarci A, Hasturk H, Van Dyke TE. Inflammatory and immune pathways in the pathogenesis of periodontal disease. Periodontology 2000. 2014;**64**(1):57-80

[3] Preshaw PM. Periodontal disease pathogenesis. In: Takei HH, Klokkevold PR, Carranza FA, editors. Carranza's Clinical Periodontology. 13th ed. Philadelphia: Elsevier Saunders; 2018. pp. 89-111

[4] Tribble GD, Lamont RJ. Bacterial invasion of epithelial cells and spreading in periodontal tissue. Periodontology 2000. 2010;**52**(1):68-83

[5] Hernández M, Dutzan N, García-Sesnich J, Abusleme L, Dezerega A, Silva N, et al. Host-pathogen interactions in progressive chronic periodontitis. Journal of Dental Research. 2011;**90**(10):1164-1170

[6] Machado V, Mesquita MF, Bernardo MA, Casal E, Proença L, Mendes JJ. IL-6 and TNF-α salivary levels according to the periodontal status in Portuguese pregnant women. PeerJ. 2018;**6**:e4710

[7] Dutzan N, Gamonal J, Silva A, Sanz M, Vernal R. Over-expression of forkhead box P3 and its association with receptor activator of nuclear factor-kappa B ligand, interleukin (IL)-17, IL-10 and transforming growth factor-beta during the progression of chronic periodontitis. Journal of Clinical Periodontology. 2009;**36**:396-403

[8] Issaranggun Na Ayuthaya B, Everts V, Pavasant P. The immunopathogenic and immunomodulatory effects of interleukin-12 in periodontal disease. European Journal of Oral Sciences. 2018;**126**:75-83

[9] Chen D, Zhang TL, Wang X. Association between polymorphisms in interleukins 4 and 13 genes and chronic periodontitis in a Han Chinese population. Biomed Research International. 2016;**2016**:8389020

[10] Shaker OG, Ghallab NA. IL-17 and IL-11 GCF levels in aggressive and chronic periodontitis patients: Relation to PCR bacterial detection. Mediators of Inflammation. 2012;**2012**:174764

[11] Pizzo G, Guiglia R, Russo L, Campisi G. Dentistry and internal medicine: From the focal infection theory to the periodontal medicine concept. European Journal Of Internal Medicine. 2010;**21**(6):496-502

[12] Tsuge A, Noda K, Nakamura Y. Early tissue reaction in the tension zone of PDL during orthodontic tooth movement. Archives of Oral Biology. 2016;**65**:17-25

[13] Liu O, Xu J, Ding G, Liu D, Fan Z, Zhang C, et al. Periodontal ligament stem cells regulate B lymphocyte function via programmed cell death protein 1. Stem Cells. 2013;**31**:1371-1382

[14] Jose JA, Somaiah S, Muddaiah S, Shetty B, Reddy G, Roopa S. A comparative evaluation of interleukin 1 beta and prostaglandin E2 with and without low-level laser therapy during en masse retraction. Contemporary Clinical Dentistry. 2018;**9**(2):267-275

[15] Meeran NA. Biological response at the cellular level within the periodontal ligament on application of orthodontic force—An update. Journal of Orthodontic Science. 2012;**1**(1):2-10

[16] Mao Y, Wang L, Zhu Y, Liu Y, Dai H, Zhou J, et al. Tension force-induced bone formation in orthodontic tooth movement via modulation of the GSK-3β/β-catenin signaling pathway. Journal of Molecular Histology. 2018;**49**(1):75-84

[17] Wise GE, King GJ. Mechanisms of tooth eruption and orthodontic tooth movement. Journal of Dental Research. 2008;**87**(5):414-434

[18] Krishnan V, Davidovitch Z. Cellular, molecular, and tissue-level reactions to orthodontic force. American Journal of Orthodontics and Dentofacial Orthopedics. 2006;**129**:469.e1-469.e32

[19] Dudic A, Kiliaridis S, Mombelli A, Giannopoulou C. Composition changes in gingival crevicular fluid during orthodontic tooth movement: Comparisons between tension and compression sides. Europrean Journal of Oral Sciences. 2006;**114**:416-422

[20] Andrade I, Silva TA, Silva GAB, Teixeira AL, Teixeira MM. The role of tumor necrosis factor type 1 in orthodontic tooth movement. Journal of Dental Research. 2007;**86**:1089-1094

[21] Basaran G, Ozer T, Kaya FA, Hamamci O. Interleukins 2, 6, and 8 levels in human gingival sulcus during orthodontic treatment. American Journal of Orthodontics and Dentofacial Orthopedics. 2006;**130**(1):7.e1-7.e6

[22] Rody WJ Jr, Wijegunasinghe M, Holliday LS, McHugh KP, Wallet SM. Immunoassay analysis of proteins in gingival crevicular fluid samples from resorbing teeth. Angle Orthodontist. 2016;**86**(2):187-192

[23] Li Y, Li M, Tan L, Huang S, Zhao L, Tang T, et al. Analysis of time-course gene expression profiles of a periodontal ligament tissue model under compression. Archives of Oral Biology. 2013;**58**(5):511-522

[24] Luppanapornlarp S, Iida J. Orthodontic force, tooth movement, and interleukin-1β. Hokkaido Journal of Dental Science. 2017;**38**:20-27

[25] Davidovitch Z. Cell biology associated with orthodontic tooth movement. In: Berkovitz BK, Moxham BJ, Newman HN, editors. The Periodontal Ligament in Health and Disease. St Louis: Mosby-Wolfe; 1995

[26] Lee TY, Lee KJ, Baik HS. Expression of IL-1beta, MMP-9 and TIMP-1 on the pressure side of gingiva under orthodontic loading. Angle Orthodontist. 2009;**79**:733-739

[27] Allgayer S, Macedo de MenezesL,RoennauLemosRM.Cytokines in crevicular fluid and orthodontic force: A systematic review. Revista Odonto Ciência. 2017;**32**(2):104-110

[28] de Oliveira Chami V, Nunes L, Capelli J Jr. Expression of cytokines in gingival crevicular fluid associated with tooth movement induced by aligners: A pilot study. Dental Press Journal of Orthodontics. 2018;**23**(5):41-46

[29] Iwasaki LR, Crouch LD, Tutor A, Gibson S, Hukmani N, Marx DB, et al. Tooth movement and cytokines in gingival crevicular fluid and whole blood in growing and adult subjects. American Journal of Orthodontics and Dentofacial Orthopedics. 2005;**128**(4):483-491

[30] Anand Kumar A, Saravanan K, Kohila K, Sathesh Kumar S. Biomarkers in orthodontic tooth movement. Review article. Dental Science. 2015;**7**(6):325-330

[31] Franchimont N, Wertz S, Malaise M. Interleukin-6: An osteotropic factor influencing bone formation. Bone. 2005;**37**(5):601-606

[32] Baggiolini M, Walz A, Kunkel SL. Neutrophil-activating

peptide-1/IL-8, a novel cytokine that activates neutrophils. Journal of Clinical Investigation. 1989;**84**:1045-1049

[33] Baumrind S, Korn EL, Boyd RL. Apical root resorption in orthodontically treated adults. American Journal of Orthodontics and Dentofacial Orthopedics. 1996;**110**:311-320

[34] Tuncer BB, Ozmeriç N, Tuncer C, Teoman I, Cakilci B, Yücel A, et al. Levels of interleukin-8 during tooth movement. Angle Orthodontist. 2005;**75**:631-636

[35] Alhashimi N, Frithiof L, Brudvik P, Bakhiet M. Orthodontic tooth movement and de novo synthesis of proinflammatory cytokines. American Journal of Orthodontics and Dentofacial Orthopedics. 2001;**119**:307-312

[36] Kohara H, Kitaura H, Yoshimatsu M, Fujimura Y, Morita Y, Eguchi T, et al. Inhibitory effect of interferon-γ on experimental tooth movement in mice. Journal of Interferon & Cytokine Research. 2012;**32**:426-431

[37] Mermut S, Bengi AO, Akin E, Kurcku M, Karacay S. Effects of interferon-gamma on bone remodeling during experimental tooth movement. Angle Orthodontist. 2007;**77**:135-141

[38] Kapoor P, Kharbanda OP, Monga N, Miglani R, Kapila S. Effect of orthodontic forces on cytokine and receptor levels in gingival crevicular fluid: A systematic review. Progress in Orthodontics. 2014;**15**:65

[39] Madureira DF, da Silva JM, Teixeira AL, Abreu MH, Pretti H, Lages EM, et al. Cytokine measurements in gingival crevicular fluid and periodontal ligament: Are they correlated? American Journal of Orthodontics and Dentofacial Orthopedics. 2015;**148**:293-301

[40] Maltha JC, van Leeuwen EJ, Dijkman G, Kuijpers-Jagtman AM. Incidence and severity of root resorption in orthodontically moved premolars in dogs. Orthodontics & Craniofacial Research. 2004;**7**(2):115-121

[41] Yamaguchi M, Yoshii M, Kasai K. Relationship between substance P and interleukin-1beta in gingival crevicular fluid during orthodontic tooth movement in adults. European Journal of Orthodontics. 2006;**28**(3):241-246

[42] Al-Ghurabi BH, Mohammed-Salih HS, Ghazi A, Saloom H. Evaluation of salivary levels of proinflammatory cytokines (IL-1α, IL-8 and GM-CSF) in adult orthodontic patients. Journal of Dental and Medical Sciences (IOSR-JDMS). 2014;**13**:75-78

[43] Nieto-Nieto N, Solano JE, Yañez-Vico R. External apical root resorption concurrent with orthodontic forces: The genetic influence. Acta Odontologica Scandinavica. 2017;**75**(4):280-287

[44] Kim N, Kadono Y, Takami M, Lee J, Lee SH, Okada F, et al. Osteoclast differentiation independent of the TRANCE-RANK-TRAF6 axis. Journal of Experimental Medicine. 2005;**202**:589-595

[45] Barbieri G, Solano P, Alarcon JA, Vernal R, Rios-Lugo J, Sanz M, et al. Biochemical markers of bone metabolism in gingival crevicular fluid during early orthodontic tooth movement. Angle Orthodontist. 2013;**83**:63-69

[46] Kunii R, Yamaguchi M. Role of interleukin-6 in orthodontically induced inflammatory root resorption in humans. Korean Journal of Orthodontics. 2013;**43**:294-301

[47] Toraldo G, Roggia C, Qian WP, Pacifici R, Weitzmann MN. IL-7 induces

bone loss in vivo by induction of receptor activator of RANKL and TNFα from T cells. Proceedings of the National Academy of Sciences. 2003;**100**:125-130

[48] Ahuja R, Almuzian M, Khan A, Pascovici D, Dalci O, Ali Darendeliler M. A preliminary investigation of short-term cytokine expression in gingival crevicular fluid secondary to high-level orthodontic forces and the associated root resorption: Case series analytical study. Progress in Orthodontics. 2017;**18**:23

[49] Takayanagi H, Sato K, Takaoka A, Taniguchi T. Interplay between interferon and other cytokine systems in bone metabolism. Immunology Reviews. 2005;**208**:181-193

[50] Walsh MC, Kim N, Kadono Y, Rho J, Lee SY, Lorenzo J, et al. Osteoimmunology: Interplay between the immune system and bone metabolism. Annual Review of Immunology. 2006;**24**:33-63

[51] Syed S, Kumuda MN, Veenadharini G, Sachidananda C, Ambica SW. Iatrogenic damage to periodontium by restorative treatment procedures: An overview. Open Dental Journal. 2015;**9**:217-222

[52] Bergman B, Hugoson A, Olsson CO. Periodontal and prosthetic conditions in patients treated with removable partial dentures and artificial crowns. Alongitudinal two-year study. Acta Odontologica Scandinavica. 1971;**29**(6):621-638

[53] Zhang ZH, Zhang XH, Zhang Y, Fang M. Detection of inflammatory factor and protease content as well as oxidative progress level in gingival crevicular fluid after fiber post and metal post repair. Journal of Hainan Medical University. 2016;**22**(14):148-151

[54] Neslihan C, Seda A, Mehmet AG, Nilgun S. The effect of restorative materials on cytokines in gingival crevicular fluid. Archives of Oral Biology. 2017;**84**:139-144

[55] Willershausen B, Kottgen C, Ernest CP. The influence of restorative materials on marginals gingiva. European Journal of Medical Reasearch. 2001;**6**(10):433-439

[56] Kosyfaki P, del Pilar Pinilla Martin M, Strub JR. Relathionship between crowns and the periodontium: A literature update. Quintessence International. 2010;**41**(2):109-126

[57] Dimova M, Doseva V. Changes in correlations between cervical crowns edge and marginal gingiva in fixed prosthodontics. Journal of IMAB—Annual Proceeding (Scientific Papers). 2006;**12**(2)

[58] Harish PV, Sonila AJ, Sirajuddin S, Gundapaneni V, Chungkham S. Iatrogenic damage to the periodontium caused by fixed prsthodontic treatment procedures. Open Dental Journal. 2015;**9**:217-222

[59] Atieh MA, Faggion CM Jr, Seymour GJ. Cytokines in patients with type 2 diabetes and chronic periodontitis: A systematic review and meta-analysis. Diabetes Research and Clinical Practice. 2014;**104**(2):e38-e45

[60] Borilova Linhartova P, Poskerova H, Tomandlova M, Bartova J, Kankova K, Fassmann A, et al. Interleukin-1 gene variability and plasma levels in Czech patients with chronic periodontitis and diabetes mellitus. International Journal of Dentistry. 2019;**2019**:6802349

[61] Acharya AB, Satyanarayan A, Thakur SL. Status of association studies linking diabetes mellitus and periodontal disease in India. International Journal of Diabetes in Developing Countries. 2010;**30**(2):69-74

[62] Longo PL, Artese HP, Rabelo MS, Kawamoto D, Foz AM, Romito GA, et al. Serum levels of inflammatory markers in type 2 diabetes patients with chronic periodontitis. Journal of Applied Oral Science. 2014;**22**(2):103-108

[63] Miranda TS, Heluy SL, Cruz DF, da Silva HDP, Feres M, Figueiredo LC, et al. The ratios of pro-inflammatory to anti-inflammatory cytokines in the serum of chronic periodontitis patients with and without type 2 diabetes and/or smoking habit. Clinical Oral Investigations. 2019;**23**(2):641-650

[64] Acharya AB, Thakur S, Muddapur MV, Kulkarni RD. Cytokine ratios in chronic periodontitis and type 2 diabetes mellitus. Diabetes & Metabolic Syndrome. 2017;**11**(4):277-278

[65] Wang S, Liu J, Zhang J, Lin J, Yang S, Yao J, et al. Glycemic control and adipokines after periodontal therapy in patients with type 2 diabetes and chronic periodontitis. Brazilian Oral Research. 2017;**31**:e90

[66] Koçak E, Sağlam M, Kayış SA, Dündar N, Kebapçılar L, Loos BG, et al. Nonsurgical periodontal therapy with/without diode laser modulates metabolic control of type 2 diabetics with periodontitis: A randomized clinical trial. Lasers in Medical Science. 2016;**31**(2):343-353

[67] Garlet GP. Destructive and protective roles of cytokines in periodontitis: A re-appraisal from host defense and tissue destruction viewpoints. Journal of Dental Research. 2010;**89**:1349-1363

[68] King GL. The role of inflammatory cytokines in diabetes and its complications. Journal of Periodontology. 2008;**79**:1527-1534

[69] Moldawer LL. Biology of proinflammatory cytokines and their antagonists. The Journal of Critical Care Medicine. 1994;**2**:S3-S7

[70] Anirudh BA, Thakurb S, Muddapurc MV, Kulkarnid RD. Cytokine ratios in chronic periodontitis and type 2 diabetes mellitus. Diabetes & Metabolic Syndrome: Clinical Research & Reviews. 2017;**11**:277-278

[71] Popescu D, Gheorghe D, Boldeanu V, Rica A. Glycemic control and interleukin-1β in periodontal patients with type 1 and 2 diabetes in a group of subjects in south-west Romania. Revista de Chimie. 2017;**68**(12):3002-3005

[72] Polak D, Shapira L. An update of the evidence for pathogenic mechanisms that may link periodontitis and diabetes. Journal of Clinical Periodontolology. 2018;**45**(2):150-166

[73] Thunell DH, Tymkiw KD, Johnson JK, Joly S, Burnell KK, Cavanaugh JE, et al. A multiplex immunoassay demonstrates reductions in gingival crevicular fluid cytokines following initial periodontal therapy. Journal of Periodontal Research. 2010;**45**(1):148-152

[74] Noh MK, Jung M, Kim SH. Assessment of IL-6, IL-8 and TNF-α levels in the gingival tissue of patients with periodontitis. Experimental and Therapeutic Medicine. 2013;**6**(3):847-851

[75] Taylor GW. Bidirectional interrelationships between diabetes and periodontal diseases: An epidemiologic perspective. Annals of Periodontology. 2001;**6**(1):99-112

[76] Huang J, Xiao Y, Xu A, Zhou Z. Neutrophils in type 1 diabetes. Journal of Diabetes Investigation. 2016;**7**(5):652-663

[77] Costa PP, Trevisan GL, Macedo GO, et al. Salivary interleukin-6, matrix

metalloproteinase-8, and osteoprotegerin in patients with periodontitis and diabetes. Journal of Periodontology. 2010;**81**(3):384-391

[78] Tomita T. Apoptosis of pancreatic β-cells in Type 1 diabetes. Bosnian Journal of Basic Medical Sciences. 2017;**17**(3):183-193

[79] Salvi GE, Franco LM, Braun TM, Lee A, Rutger Persson G, Lang NP, et al. Pro-inflammatory biomarkers during experimental gingivitis in patients with type 1 diabetes mellitus: A proof-of-concept study. Journal of Clinical Periodontology. 2010;**37**(1):9-16

[80] Salvi GE, Yalda B, Collins JG, Jones BH, Smith FW, Arnold RR, et al. Inflammatory mediator response as a potential risk marker for periodontal diseases in insulin-dependent diabetes mellitus patients. Journal of Periodontology. 1997;**68**:127-135

[81] Pacios S, Kang J, Galicia J, Gluck K, Patel H, Ovaydi-Mandel A, et al. Diabetes aggravates periodontitis by limiting repair through enhanced inflammation. FASEB Journal. 2012;**26**(4):1423-1430

[82] Duque C, João MF, Camargo GA, et al. Microbiological, lipid and immunological profiles in children with gingivitis and type 1 diabetes mellitus. Journal of Applied Oral Science. 2017;**25**(2):217-226

[83] Lauterbach MA, Wunderlich FT. Macrophage function in obesity-induced inflammation and insulin resistance. Pflügers Archiv. 2017;**469**(3):385-396

[84] Schenkein HA, Loos BG. Inflammatory mechanisms linking periodontal diseases to cardiovascular diseases. Journal of Clinical Periodontology. 2013;**40**:S51-S69

[85] Nguyen CM, Kim JW, Quan VH, Nguyen BH, Tran SD. Periodontal associations in cardiovascular diseases: The latest evidence and understanding. Journal of Oral Biology and Craniofacial Research. 2015;**5**(3):203-206

[86] Berlin-Broner Y, Febbraio M, Levin L. Association between apical periodontitis and cardiovascular diseases: A systematic review of the literature. International Endodontic Journal. 2017;**50**(9):847-859

[87] El Kholy K, Genco RJ, Van Dyke TE. Oral infections and cardiovascular disease. Trends in Endocrinology & Metabolism. 2015;**26**(6):315-321

[88] Fatkhullina AR, Peshkova IO, Koltsova EK. The role of cytokines in the development of atherosclerosis. Biochemistry (Moscow). 2016;**81**(11):1358-1370

[89] Ziebolz D, Jahn C, Pegel J, Semper-Pinnecke E, Mausberg RF, Waldmann-Beushausen R, et al. Periodontal bacteria DNA findings in human cardiac tissue—Is there a link of periodontitis to heart valve disease? International Journal of Cardiology. 2018;**251**:74-79

[90] Mozos I, Malainer C, Horbańczuk J, Gug C, Stoian D, Luca CT, et al. Inflammatory markers for arterial stiffness in cardiovascular diseases. Frontiers in Immunology. 2017;**8**:1058

[91] Chistiakov DA, Orekhov AN, Bobryshev YV. Links between atherosclerotic and periodontal disease. Experimental and Molecular Pathology. 2016;**100**(1):220-235

[92] Amar S, Al-Hashemi J. Periodontal innate immune mechanisms relevant to atherosclerosis. In: Gavins NEF, Karen Y, editors. Vascular Responses to Pathogens. Cambridge: Elsevier; 2016. pp. 75-85

[93] Nguyen TT, Wu KY, Leclerc M, Pham HM, Tran SD. Cardiovascular

diseases and periodontal disease. Current Oral Health Reports. 2018;**5**(1):13-18

[94] Hansson GK, Hermansson A. The immune system in atherosclerosis. Nature Immunology. 2011;**12**(3):204

[95] Choy E. Understanding the dynamics: Pathways involved in the pathogenesis of rheumatoid arthritis. Rheumatology (Oxford, England). 2012;**51**(5):3-11

[96] Kempeni J. Update on D2E7: A fully human anti-tumour necrosis factor α monoclonal antibody. Annals of the Rheumatic Diseases. 2000;**59**(S1):144-145

[97] Barrera P, Joosten LA, den Broeder AA, van de Putte LB, vanRiel PL, van den Berg WB, et al. Effects of treatment with a fullyhuman anti-tumour necrosis factor α monoclonal antibody on the local and systemic homeostasis of interleukin 1 and TNF α in patients with rheumatoid arthritis. Annals of the Rheumatic Diseases. 2001;**60**:660-669

[98] Spadaro A, Rinaldi T, Riccieri V, Taccari E, Valesini G. Interleukin-13 in autoimmune rheumatic diseases: Relationship with the autoantibody profile. Clinical and Experimental Rheumatology. 2002;**20**(2):213-216

[99] Silosi I, Cojocaru M, Foia L, Boldeanu MV, Petrescu F, Surlin P, et al. Significance of circulating and crevicular matrix metalloproteinase-9 in rheumatoid arthritis-chronic periodontitis association. Journal of Immunology Research. 2015;**2015**:218060

[100] Koziel J, Mydel P, Potempa J. The link between periodontal disease and rheumatoid arthritis: An updated review. Current Rheumatology Reports. 2014;**16**:408

[101] Kaur S, White S, Bartold PM. Periodontal disease and rheumatoid arthritis: A systematic review. Journal of Dental Research. 2013;**92**(5):399-408

[102] Ogrendik M. Rheumatoid arthritis is linked to oral bacteria: Etiological association. Modern Rheumatology. 2009;**19**:453-456

[103] Detert J, Pischon N, Burmester GR. The association between rheumatoid arthritis and periodontal disease. Arthritis Research & Therapy. 2010;**218**(1-7):12

[104] Sarhan MA, Pham TN, Chen AY, Michalak TI. Hepatitis C virus infection of human T lymphocytes is mediated by CD5. Journal of Virology. 2012;**86**(7):3723-3735

[105] Chen SL, Morgan TR. The natural history of hepatitis C virus (HCV) infection. International Journal of Medical Sciences. 2006;**3**(2):47

[106] Bataller R, Brenner DA. Liver fibrosis. The Journal of Clinical Investigation. 2005;**115**(2):209-218

[107] Capone F, Guerriero E, Colonna G, Maio P, Mangia A, Castello G, et al. Cytokinome profile evaluation in patients with hepatitis C virus infection. World Journal of Gastroenterology. 2014;**20**(28):9261-9269

[108] Gheorghe DN, Foia L, Toma V, Surdu A, Herascu E, Popescu DM, et al. Hepatitis C infection and periodontal disease: Is there a common immunological link? Journal of Immunology Research. 2018;**2018**:8720101

[109] Barros SP, Williams R, Offenbacher S, Morelli T. Gingival crevicular fluid as a source of biomarkers for periodontitis. Periodontology 2000. 2016;**70**(1):53-64

[110] Akram Z, Abduljabbar T, Hassan A, Ibrahim M, Javed F, Vohra F. Cytokine profile in chronic periodontitis patients with and without obesity: A systematic review and meta-analysis. Disease Markers. 2016;**2016**:PMC5067315

[111] Majeed Z, Philip K, Alabsi AM, Pushparajan S, Swaminathan D. Identification of gingival crevicular fluid sampling, analytical methods, and oral biomarkers for the diagnosis and monitoring of periodontal diseases: A systematic review. Disease Markers. 2016;**2016**:1804727

[112] Sharma A, Chakraborti A, Das A, Dhiman RK, Chawla Y. Elevation of interleukin-18 in chronic hepatitis C: Implications for hepatitis C virus pathogenesis. Immunology. 2009;**128**(1pt2):e514-e522

[113] Wang J, Zhao P, Guo H, Sun X, Jiang Z, Xu L, et al. Serum IL-33 levels are associated with liver damage in patients with chronic hepatitis C. Mediators of Inflammation. 2012;**2012**:819636. 7 p

[114] Nishitsuji H, Funami K, Shimizu Y, Ujino S, Sugiyama K, Seya T, et al. Hepatitis C virus infection induces inflammatory cytokines and chemokines mediated by the cross talk between hepatocytes and stellate cells. Journal of Virology. 2013;**87**(14):8169-8178

[115] Younossi Z, Anstee QM, Marietti M, Hardy T, Henry L, Eslam M, et al. Global burden of NAFLD and NASH: Trends, predictions, risk factors and prevention. Nature Reviews Gastroenterology & Hepatology. 2018;**15**(1):11

[116] Bedogni G, Miglioli L, Masutti F, Tiribelli C, Marchesini G, Bellentani S. Prevalence of and risk factors for nonalcoholic fatty liver disease: The Dionysos nutrition and liver study. Hepatology. 2005;**42**(1):44-52

[117] Bellentani S. The epidemiology of non-alcoholic fatty liver disease. Liver International. 2017;**37**:81-84

[118] Abdelmalek MF, Diehl AM. Nonalcoholic fatty liver disease as a complication of insulin resistance. Medical Clinics of North America. 2007;**91**(6):1125-1149

[119] Seo YY, Cho YK, Bae JC, Seo MH, Park SE, Rhee EJ, et al. Tumor necrosis factor-α as a predictor for the development of nonalcoholic fatty liver disease: A 4-year follow-up study. Endocrinology and Metabolism. 2013;**28**(1):41-45

[120] Singh P, Gupta ND, Bey A, Khan S. Salivary TNF-alpha: A potential marker of periodontal destruction. Journal of Indian Society of Periodontology. 2014;**18**(3):306-310

[121] Stadler AF, Angst PD, Arce RM, Gomes SC, Oppermann RV, Susin C. Gingival crevicular fluid levels of cytokines/chemokines in chronic periodontitis: A meta-analysis. Journal of Clinical Periodontology. 2016;**43**(9):727-745

[122] Algate K, Haynes DR, Bartold PM, Crotti TN, Cantley MD. The effects of tumour necrosis factor-α on bone cells involved in periodontal alveolar bone loss; osteoclasts, osteoblasts and osteocytes. Journal of Periodontal Research. 2016;**51**(5):549-566

[123] Arsenie C, Sandulescu DL, Popescu DM, Gheorghe DN, Martu A, Foia L, et al. Periodontal changes and the non-alcoholic fatty liver disease. International Journal of Medical Dentistry. 2018;**22**(3):280-287

[124] Han P, Sun D, Yang J. Interaction between periodontitis and liver diseases. Biomedical Reports. 2016;**5**(3):267-276

[125] Kuraji R, Ito H, Fujita M, Ishiguro H, Hashimoto S, Numabe Y. Porphyromonas gingivalis induced periodontitis exacerbates progression of non-alcoholic steatohepatitis in rats. Clinical and Experimental Dental Research. 2016;**2**(3):216-225

[126] Duarte PM, Bastos MF, Fermiano D, Rabelo CC, Perez-Chaparro PJ, Figueiredo LC, et al. Do subjects with aggressive and chronic periodontitis exhibit a different cytokine/chemokine profile in the gingival crevicular fluid? A systematic review. Journal of Periodontal Research. 2015;**50**(1):18-27

[127] Goral V, Atayan Y, Kaplan A. The relation between pathogenesis of liver cirrhosis, hepatic encephalopathy and serum cytokine levels: What is the role of tumor necrosis factor α? Hepato-Gastroenterology. 2011;**58**(107-108):943-948

[128] Alazawi W, Bernabe E, Tai D, Janicki T, Kemos P, Samsuddin S, et al. Periodontitis is associated with significant hepatic fibrosis in patients with non-alcoholic fatty liver disease. PLoS One. 2017;**12**(12):e0185902

[129] Costantini S, Capone F, Guerriero E, Maio P, Colonna G, Castello G. Serum cytokine levels as putative prognostic markers in the progression of chronic HCV hepatitis to cirrhosis. European Cytokine Network. 2010;**21**(4):251-256

[130] Cardoso EM, Reis C, Manzanares-Céspedes MC. Chronic periodontitis, inflammatory cytokines, and interrelationship with other chronic diseases. Postgraduate Medicine. 2018;**130**(1):98-104

[131] Gheorghe DN, Camen A, Foia L, Solomon S, Toma V, Mateescu OG, et al. Histologic and immunohistochemical assessment of gingival Tissue's changes induced by periodontal disease in association with chronic hepatitis C. Medical-Surgical Journal-Revista Medico-Chirurgicala. 2018;**122**(4):789-797

[132] Șurlin P, Camen A, Stratul SI, Roman A, Gheorghe DN, Herăscu E, et al. Optical coherence tomography assessment of gingival epithelium inflammatory status in periodontal—Systemic affected patients. Annals of Anatomy-Anatomischer Anzeiger. 2018;**219**:51-56

[133] Bokhari SA, Khan AA. Growing burden of noncommunicable diseases: The contributory role of oral diseases, Eastern Mediterranean Region perspective. East Mediterranean Health Journal. 2009;**15**(4):1011-1020

[134] Levin A, Tonelli M, Bonventre J. Global kidney health 2017 and beyond: A roadmap for closing gaps in care, research, and policy. Lancet. 2017;**390**:1888-1917

[135] Seikaly MG, Loleh S, Rosenblum A, Browne R. Validation of the Center for Medicare and Medicaid Services algorithm for eligibility for dialysis. Pediatric Nephrology. 2004;**19**(8):893-897

[136] Neild GH. Life expectancy with chronic kidney disease: An educational review. Pediatric Nephrology. 2017;**32**(2):243-248

[137] Eustace JA, Astor B, Muntner PM, et al. Prevalence of acidosis and inflammation and their association with low serumalbumin in chronic kidney disease. Kidney International. 2004;**65**:1031-1040

[138] Carrero JJ, Stenvinkel P. Inflammation in end-stage renal disease—What have we learned in 10 years? Seminars in Dialysis. 2010;**23**:498-509

[139] Brito F, Ameida S, Figueredo CM, et al. Extent and severity of chronic

periodontitis in chronic kidney disease patients. Journal of Periodontology Research. 2012;**47**:426-430

[140] Spoto B, Mattace-Raso F, Sijbrands EL. Association of IL-6 and a functional polymorphism in the IL-6 gene with cardiovascular events in patients with CKD. Clinical Journal of American Society of Nephrology. 2015;**10**(2):232-240

[141] Lam WKC. Inflammation, cytokines and chemokines in chronic kidney disease. EJIFCC. 2009;**20**(1):12-20

[142] Lima SM, Otoni A, Sabino Ade P. Inflammation, neoangiogenesis and fibrosis in peritoneal dialysis. Clinica Chimica Acta. 2013;**421**:46-50

[143] Sincar C, Solomon S, Rudnic I, Martu I, Martu S. The efects of periodontal therapy on biochemical inflammatory markers on renal dysfunction patiens. Romanian Journal of Oral Rehabilitation. 2013;**5**(1):71-76

[144] Sincar C, Solomon S, Rudnic I, Martu S. Evaluation of odonto-periodontal parameters in patients with chronic renal insufficiency treated by hemodialysis Int. International Journal of Medical Dentistry. 2012;**5**(2):214-219

[145] Veisa G, Tasmoc A, Nistor I, Segall L, Siriopol D, Solomon SM, et al. The impact of periodontal disease on physical and psychological domains in long-term hemodialysis patients: A cross-sectional study. International Urology and Nephrology. 2017;**49**(7):1261-1266